QH585
NUN
WITHDRAWN
FROM STOCK
QMUL LIBRARY

Flow Cytometry for Research Scientists: Principles and Applications

Rafael Nunez M.D. M.Sc.

Institute of Virology
University of Zurich
Winterthurerstr. 266a
8057 Zurich, Switzerland
rafaeln@vetvir.unizh.ch

Copyright © 2001
Horizon Press
32 Hewitts Lane
Wymondham
Norfolk NR18 0JA
U.K.

www.horizonpress.com

British Library Cataloguing-in-Publication Data

A catalogue record for this book is available from the British Library

ISBN: 1-898486-26-3

Description or mention of instrumentation, software, or other products in this book does not imply endorsement by the author or publisher. The author and publisher do not assume responsibility for the validity of any products or procedures mentioned or described in this book or for the consequences of their use.

All rights reserved. No part of this publication may be reproduced, stored in a retrieval system, or transmitted, in any form or by any means, electronic, mechanical, photocopying, recording or otherwise, without the prior permission of the publisher. No claim to original U.S. Government works.

Printed and bound in Great Britain
by Biddles Ltd Guildford and King's Lynn

Contents

Contents	iii
Acknowledgements	vi
Chapter 1 Introduction to the Field of Cytometry and its Importance in Biomedicine	1
Chapter 2 Overview of Flow Cytometry Instruments	5
Chapter 3 Flow Cytometric Data Analysis. CellQuest Software	17
Chapter 4 DNA Analysis. DNA Measurement and Cell Cycle Analysis	29
Chapter 5 Molecular Cytometry	37
Chapter 6 Surface Staining and Immunophenotyping Using Multicolor Analysis	41
Chapter 7 Handling of Samples: Biosafety	53
Chapter 8 Intracellular Antigens. Cytokines and the Study of Viral Antigens	55
Chapter 9 Flow Cytometric Assessment of Sperm Quality and Cell Cycle Analysis	57

Chapter 10   63
Flow Cytometric Assessment of Allopurinol Susceptibility
in *Leishmania infantum* Promastigote

Chapter 11   67
Flow Cytometric Assessment of Transduction Efficiency
and Vector Cytotoxicity of HSV-1 Amplicon Vectors

Chapter 12   71
Immunophenotyping of DC by Flow Cytometry
and Description of Diverse Functional Studies

Chapter 13   83
Summary

Chapter 14   89
Protocols

Appendix 1   95
Commercial Resources for Flow Cytometry on the Web

Appendix 2   97
Abbreviations used in Flow Cytometry

Index   99

Books of Related Interest

Gene Cloning and Analysis: Current Innovations	1997
An Introduction to Molecular Biology	1997
Genetic Engineering with PCR	1998
Prions: Molecular and Cellular Biology	1999
Probiotics: A Critical Review	1999
Peptide Nucleic Acids: Protocols and Applications	1999
Intracellular Ribozyme Applications: Principles and Protocols	1999
NMR in Microbiology: Theory and Applications	2000
Molecular Marine Microbiology	2000
Oral Bacterial Ecology: The Molecular Basis	2000
Prokaryotic Nitrogen Fixation: A Model System	2000
Cold Shock Response and Adaptation	2000
Development of Novel Antimicrobial Agents: Emerging Strategies	2001
***H. pylori*: Molecular and Cellular Biology**	2001
Environmental Molecular Microbiology	2001
Genomes and Databases on the Internet	2001
The Internet for Cell and Molecular Biologists	2001

For further information on these books contact:

Horizon Scientific Press, P.O. Box 1, Wymondham, Norfolk, NR18 0EH, U.K.
Tel: +44(0)1953-601106. Fax: +44(0)1953-603068. Email: mail@horizonpress.com

Our Web site has details of all our books including full chapter abstracts, book reviews, and ordering information:

www.horizonpress.com

Acknowledgements

I would like to thank Professor M. Ackermann for support, encouragement and advice. Also I thank Dr. P. D. M. Suter for support and helpful suggestions, and the personnel of the Institute of Virology for valuable comments and suggestions. I also thank my colleagues for their friendship and scientific support, and in particular Steven Z. Merlin (HSS New York), Mario Roederer (Stanford University), Joseph Martinez (CDC Atlanta), Tom Just (DAKO-Denmark) and Michael G. Ormerod for kindly providing reprints and materials.

I would like to thank Steven Z. Merlin (HSS New York) for input to chapters 1, 2, 3, 4, 5, and 7, and Dr. C. Nunez M.D. (MD Anderson, Houston) for input to chapters 6 and 12.

1

Introduction to the Field of Cytometry and its Importance in Biomedicine

Flow cytometry is a methodology for determining and quantitating cellular features, organelles or cell structural components primarily by both optical and electronic means. Although it measures one cell at a time, the newest equipment is able to process up to several hundred thousand cells in a few seconds. Flow cytometry can be used to count and even distinguish cells of different types in a mixture by quantitating their structural features. Therefore, flow cytometry has great advantages compared to traditional microscopy since it permits the analysis of a greater number of cells in a fraction of the time. In addition cell sorting with flow cytometers has been a powerful tool for diverse fields in biomedical research and clinical applications.

What is Flow Cytometry and its importance?
Since the early 1970's, flow cytometers that do not employ fluorescence have been commercially available. They were initially used for complete blood cell counts in clinical laboratories. Their ease of handling and reliability of results increased and popularized their use. The newest and most versatile research instruments employ fluorescence; these are named flow cytofluorometers. The world-wide utilization of flow cytometry is demonstrated by the occurance of flow cytometric data in almost any issue of a scientific journal concerned with cell biology. In addition, a large percentage of research papers in the fields of biomedical sciences and immunology report flow cytometric data. About 43,200 citations containing flow cytometry information has been compiled to date by MEDLINE (July, 2000).

Flow cytometers are widely found in all leading biomedical research institutions and universities where they are used for performing tasks that require analytical precision and high throughput. In addition, flow cytometers have a key role in hospital and medical centers worldwide, where they are widely used for diagnosis as well as research. There are several thousand flow cytofluorometers in clinical use worldwide. The major diagnostic applications being ploidy, cell cycle and surface analysis of cancers. They are also of use in the study of surface markers of lymphomas and leukemias which are of diagnostic and prognostic value. Flow cytometry also has been the method of choice for monitoring the progression of AIDS and the response to treatment by measuring CD4 lymphocyte levels in the blood. Less expensive alternative technologies are not yet available for performing such tasks in clinical and research settings. In addition, sorting and high speed sorting are becoming increasingly important in the performance of research, clinical trials, clinical applications and teaching.

How Does It Work?

The cells prepared in a monodisperse (single cell) suspension may be alive or fixed at the time of measurement. They are passed through a chamber as single cells. The fine stream containing the cellular suspension is passed through the chamber as a continuous flow. The cells inside the chamber are excited by the beam of the laser(s) light. Each cell scatters some of the laser light, as well as emitting fluorescent light following excitation by the laser. The signal of the scattered light and the signal of the emitted fluorescence are collected for analysis. The cytometer typically measures several parameters simultaneously for each cell:

i) Forward scatter intensity (FSC) is approximately proportional to cell diameter.
ii) Side scatter intensity (SSC) or orthogonal (90 degrees) scatter is approximately proportional to the quantity of granular structures within the cell.
iii) Fluorescence intensities are measured at several wavelengths.

FSS alone is often quite useful. It is commonly used to exclude dead cells, cell aggregates, and cell debris from the fluorescence data. It is sufficient to distinguish lymphocytes from monocytes or from granulocytes in blood leukocyte samples. Side scatter has been used in our laboratory to assess granularity of living cells such as Dendritic cells.

Fluorescence intensities are typically measured at several different wavelengths simultaneously for each cell. Fluorescent probes, generally coupled to antibodies, are used to report the quantities of specific components of the cells. Fluorescent antibodies are often used to report the densities of specific surface receptors, and thus to distinguish subpopulations of differentiated cell types, including cells expressing a transgene or cells expressing a unique marker.

The binding to surface receptors of viruses, or diverse proteins such as hormones, can also be measured by making them fluorescent. Intracellular components can also be reported by using fluorescent probes. For example, measurement of total DNA per cell will allow cell cycle analysis. Also, cytometric determination of newly synthesized DNA or identification of specific nucleotide sequences in DNA or mRNA can be accomplished. Filamentous actin or any structure for which an antibody is available can be determined by cytometry.

Flow cytometry can also be used to measure rapid changes in intracellular free calcium, membrane potential, pH, or free fatty acids. In addition, flow cytometry can be used for measurement of the metabolic status of cellular membranes and to study mitochondria functionality.

Flow cytometers involve sophisticated handling of fluids and pressure, complex laser beams and optics, very sensitive electronic detectors, analogue to digital converters, and high capability computers. The optics deliver laser light inside the chamber focused to a beam a few cell diameters in size. The fluidics hydrodynamically focus the cell stream within a margin of a small fraction of a cell diameter, and, even in sorters, break the stream into uniform-sized droplets in order to separate individual cells. The electronics quantitate the faint flashes of scattered and fluorescent light, and, under computer control, select the electrically charged droplets containing cells of interest so that they can be deflected into a separate test tube or culture wells. The computer is able to record data for thousands of cells per sample, and displays the data graphically.

The sorting applications have evolved from obtaining a few cells to complex applications the selection of cells for single cell analysis and/or the bulk collection of cells with unique characteristics that will be analyzed or transferred into a receptical.

General References
1. Robinson, J.P. 1998. Current Protocols in Cytometry. John Wiley & Sons, Inc. New York.
2. Coligan, J.E. 1998. Current Protocols in Immunology. John Wiley & Sons, Inc. New York.
3. CellQuest software reference manual. 1997. Becton Dickinson immunocytometry systems.
4. Watson, J. 1998. Purdue Cytometry CD-ROM. Volume 4, Purdue University Cytometry Laboratories. http://www.cyto.purdue.edu

5. Ormerod, M. 1999. Flow cytometry. Bios Scientific, ISBN: 185996107X.
6. Ormerod, M. Data Analysis in Flow Cytometry: A Practical Approach. A CD-ROM.
   http://ourworld.compuserve.com/homepages/Michael_Ormerod.
7. Steinkamp, J. 1984. Flow cytometry. Rev. Sci. Instrum. 55: 1375-1400.
8. The Art of Fluorescent Activated Cell Sorting: Research Notes from Becton Dickinson Order number 23-2137-00.
9. Givan, AL. 1992. Flow Cytometry: First Principles. Wiley-Liss, New York.
10. Bagwell CB. 1993. Theoretical Aspects of Data Analysis. In: Clinical Flow Cytometry Principles and Application. Ed Bauer KD, Duque RE and Shankey TV. Pages 41-61.

2

Overview of Flow Cytometry Instruments

The standard benchtop flow cytometer is very similar to a hematology cell counter. In fact, flow cytometers can trace their origins back to the early hematology counters. Unlike their earlier counterparts that used electronic impedance to measure particles in a fluid stream, today's modern flow cytometers use an illuminating light source usually consisting of a laser or arc lamp. The majority of instrument manufacturers employ an air-cooled argon gas laser emitting a monochromatic beam of light fixed at 488 nm at 15 mW of power. As particles or cells flow in single file past the intersection of the light beam, light is scattered in various directions. If there a fluorochrome labeled monoclonal antibody associated with the cell, it becomes excited by the laser and a fluorescent emission results. The resulting signals are processed to gather information about the relative size of the cell (forward light scatter-FSC), its shape or internal complexity (side light scatter-SSC) as well as a diversity of cellular structures and antigens (fluorescence).

The cytometer itself is set up and monitored routinely with a quality control program utilizing a series of unlabeled and fluorescently labeled calibration particles as a reference check on instrument alignment and sensitivity performance. Therefore, every cytometric procedure performed is a reflection of the quality of the instrument, sample preparation, and the operator, and ultimately reflects on the institution.

FACSCalibur™
The FACSCalibur uses an air-cooled argon gas laser regulated at 15 mW of power and a fixed wavelength emission of 488 nm. The instrument is capable of detecting six parameters: Forward Scatter (FSC), side scatter (SSC) and three fluorescent emissions (green, yellow-orange and red) utilizing the first laser. A smaller diode laser emitting red light at 635 nm is used to excite compounds that fluoresce above 650 nm and are detected with the fourth

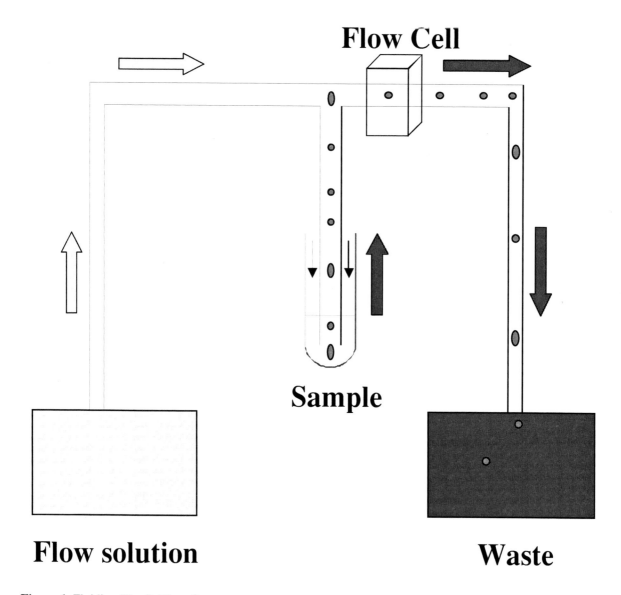

Figure 1. Fluidics. The fluidics of a cytometer.

fluorescent detector (photomultiplier tube). Figures 1-3 illustrate the general system design and component layout of a flow cytometer. Figure 1 iilustrates the fluidics of a cytometer. Figure 2 shows the optics of a cytometer with two lasers and the capability to detect up to six parameters. Figure 3 demonstrates the process of acquisition and analysis of a sample and the capabilities to save the data on disks as well as to analyze the data and produce printouts and/or slides.

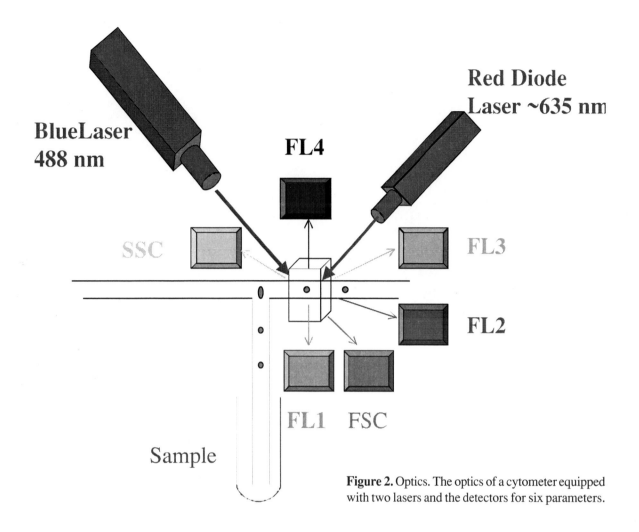

Figure 2. Optics. The optics of a cytometer equipped with two lasers and the detectors for six parameters.

There is a large number of fluorogenic dyes available that can be covalently coupled to antibodies or used for intracellular staining of various cellular constituents. The most widely used fluorochrome for immunophentyping is fluorescein isothyocyanate (FITC) and is detected in the green channel or FL1. Phycoerythrin (PE) emits in the yellow-orange region and is detected in FL2, while Propidium Iodide is a DNA specific dye that can be measured in either the orange or red channel (FL2 or FL3). Additional dyes like PerCP, TriColor and Red613 can be excited by 488 nm and emit in the red wavelengths.

FACSComp™ software is a software application for automating the setup and electronic compensation necessary when using multiple fluorochromes. FACSComp is used in conjunction with Calibrite™ beads which are polystyrene beads impregnated with the appropriate fluorochrome. The

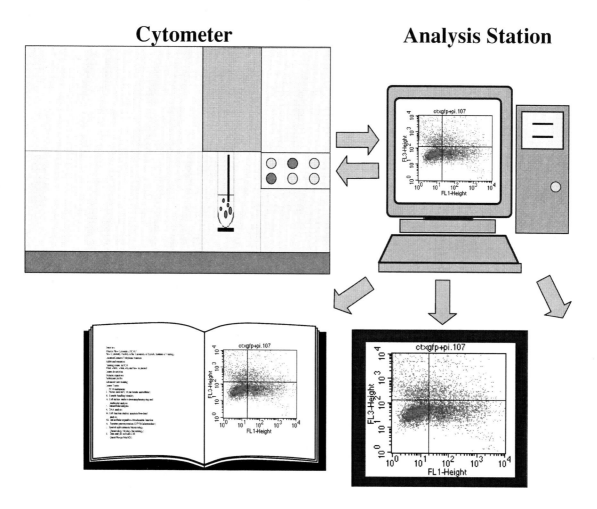

Figure 3. Acquisition and analysis. The process of acquisition and analysis of a sample and the capability to save the data on disks as well as to generate print outs and/or design slides.

software informs the user when to insert the required tube, the compensation and sensitivity being performed automatically. Results are displayed on the video monitor, saved by the system, and a printout is available as a record for quality control purposes.

Unlike the traditional benchtop cell analyzer, the FACSCalibur can separate cells into different collection containers based on their physical properties. This process of selection is the basis for cell sorting (see below). When only analyzing cells, samples are consumed and discarded. The instrument is designed with a closed fluid system, thereby making it ideally suited for working with potentially biohazardous samples, *i.e.*, human blood. Working

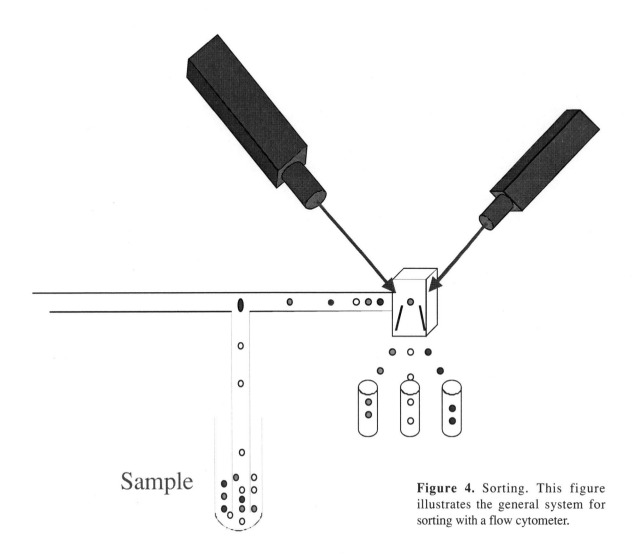

Figure 4. Sorting. This figure illustrates the general system for sorting with a flow cytometer.

with blood samples must be authorized and the appropriate precautions implemented.

The FACSCalibur can analyze cell suspensions at the rate of several hundred cells per second. Ideally, each sample to be analyzed should contain a cellular concentration of between $10^5 - 1 \times 10^6$. Typically, investigators acquire 5,000 to 10,000 cells per sample file; for sperm analysis up to 100,000 cells could be collected. Larger data files have greater storage requirements and generally require storage media such as Magneto-optical (MO) disks. Normally, data are first saved directly to the hard disk of the PowerPC Macintosh computer. After backing up to a 1.44 MB floppy or 230 MB rewritable optical disks, the data can be transferred to a FACS™ analysis data station. The off-line work station should contain a copy of the CellQuest™ data analysis software. Moreover, excellent public domain software for the PC (WinMDI™) and commercially available programs such as FlowJo™ for the Macintosh are available from sites on the internet.

The FACSCalibur is easy to use, and in our facility, the instrument is operated by the individual researcher who has participated and successfully completed the training course. The facility maintains a schedule sheet for reserving analysis time on a first-come basis. A block of time can be reserved in advance for sample acquisition or data analysis.

For additional information on the FACSCalibur instrument, reagents and software, BDIS maintains a web site at: http://www.bdfacs.com/literature/brochures/Caliburimages/CaliburBro.gif

BD LSR
The BD LSR™ is the newest instrument in the flow cytometry line. Introduced in August 1999, this 8 parameter (6 color and two scatter measurements) is a dual laser, benchtop instrument offering simultaneous 488 nm and UV excitation. It features an air-cooled 8 mW, 325 nm He-Cd laser in addition to the 15 mW, 488 nm argon-ion laser. Building upon the ease of use of the FACSCalibur, the BD LSR simplifies routine UV applications from the more complex cell sorter instruments to a high performance, compact benchtop flow cytometer. The instrument offers software instrument control, push-button fluidics and fine adjustment sample flow rate control. The optical system is alignment-free and permits the use of UV-excited dyes such as Hoechst 33342, DAPI and Indo-1.

Characterisitics of a cell sorting instrument and potential applications
A cell sorting instrument is designed first and foremost as a very high-speed sorter and analyzer. It offers the sort performance needed for the quantification and/or isolation of rare cell populations with high recovery and purity. The design specification for a cell sorting instrument includes speeds at or above 25,000 cells per second, and performance approximately 50-100 times faster than existing instruments. Sorted cells must be viable with purity and recovery better than 95 percent at any speed.

Moreover, current performance specifications for a sell sorting instrument include a speed of 50,000 cells per second, with purity and recovery greater than 99 percent at that speed.

A high speed and sensible cell sorting instrument equipped with a single-cell sorting module permits sterile cloning of single cells and allows sorting of single cells directly into tubes for subsequent PCR enzymatic amplification of DNA or RNA from sorted cells. Also on-line are special home-made, high-resolution time-of-flight sizing systems capable of sizing cells and subcellular organelles as small as 0.3 microns in length or diameter, and a multiparameter high-speed "rare-event" analysis system ("HIGHSPEED") capable of analyzing cells at rates in excess of 100,000 cells per second allowing analysis of rare cell subpopulations as small as 0.0001 percent with sample rates of more than 10^9 cells per hour.

FACSCalibur Sorting option:
As previously mentioned, the FACSCalibur is equipped with the optional sorting module. The system employs a mechanical catcher tube assembly within the closed fluid system to divert cells to the collection tubes. This method produces no biohazardous aerosols, is ideal for pathogenic samples, but is considerably slower in sorting. For sorting of non-biohazardous samples where the sorting criteria cannot be met on the FACSCalibur, a larger multi-parameter jet-in-air cell sorter should be used.

The FACSCalibur can acquire data and sort cells. While acquisition can be performed at rates over 1,000 events per second, the sorting of cells is limited to approximately 300 per second. Non hazardous living cells can be sorted, and may be recovered in gnotobiotic ("sterile") form for subsequent *in vitro* functional studies.

Although sample acquisition on the FACSCalibur is performed by the researcher, sorting of cells is a more complex procedure requiring precise alignment. Because of its open design, which permits great flexibility in configuration, its delicate components are much more easily damaged by inexperienced operators, and quite expensive to replace. Therefore operation of the FACSCalibur for sorting purposes is restricted to trained operators.

FACS Vantage™
The Becton-Dickinson FACS Vantage Cell Sorter instrument is capable of sorting cells at a much higher speed than the FACSCalibur. Unlike the Calibur which has an enclosed sorting system, the FACS Vantage uses a jet-in-air system to achieve greater sort speeds. With robust analytical capabilities and flexibility in changing the optical configuration and adding of lasers, a wider range of fluorochromes can be used for analysis and sorting applications (Figure 4).

Beckman Coulter
Beckman Coulter manufactures the COULTER® EPICS® XL/XL-MCL FlowCytometry System for bench top operations and the "EPICS® ALTRA™ and HyPerSort™ High-Performance Cell Sorting System with capability for high speed sorting applications.

Coulter® EPICS® XL/XL-MCL FlowCytometry System
The Coulter® EPICS® XL Flow Cytometer combines the analytical power of a research cytometer into a compact, ergonomically designed clinical analyzer. The XL System features the capability to analyze up to four colors of immunofluorescence from a single air cooled laser. Other multi-color applications include: multiparametric DNA analysis, platelet studies, reticulocyte enumeration, cell biology/functional studies as well as a broad range of research applications.

For research applications, the user has the ability to change the optical filters for each fluorescent measurement channel. For high volume users, the XL-MCL System offers walk-away sample handling with the Multi Carousel Loader (MCL). This device incorporates positive bar-code identification and vortex mixing prior to sample aspiration.

The XL uses Digital Signal Processing (DSP) for reliable linearity and drift-free amplification and compensation. Multi-fluorochrome compensation is achieved by a digital compensation matrix. The new XL SYSTEM II™ Software fully automates instrument set-up and compensation for 2, 3 and 4-color applications. Optional features in the software include tetraONE, the 4-color, one tube, automatic analysis algorithm and reticONE, an analysis algorithm for reticulocyte analysis.

EPICS®ALTRA™ and HyPerSort™ High-Performance Cell Sorting System
The EPICS ALTRA flow cytometer represents a new generation of cell sorters from Beckman Coulter. This high-performance instrument is a powerful tool for innovative research in immunobiology, cell physiology, molecular biology, genetics, microbiology, water quality and plant cell analysis. High fluorescence sensitivity with quartz SortSense™ flow cell design, for sense-in-quartz, jet-in-air operation. The system features the capability to analyze and sort on up to 6 colors simultaneously while performing complex multi-parameter applications. These include DNA cell-cycle analysis, quantification, functional studies, chromosome enumeration and physiological measurements.

The standard ALTRA System allows rapid separation of large numbers of specific cell populations with high purity, recovery and yield. Sorting at rates up to 10,000 cells per second (with air-cooled lasers) and up to 15,000 cells per second (with water-cooled lasers) is possible. The system features eight sort modes. Four modes optimize between purity and yield for two sorted populations, including a special enrichment mode. Three modes feature the new ALTRASort™ mixed-mode that isolates a high purity population to the left and captures the remainder of the same population to the right. This is ideal when working with rare populations and/or limited sample volume.

The AccuSort™ mode is for accurate counting when sorting individual cells for cloning with the Autoclone® sorting option or for sort matrix verification. The optional HyPerSort™ System incorporates new sorting technology that provides true high speed sorting, while maintaining sensitivity with the SortSense flow cell. HyPerSort technology uses high pressure and high frequency to maximize purity, recovery and yield at data rates up to 30,000 cells per second.

This user-configurable optical platform permits choosing from an extensive range of laser options, allowing almost any combination of excitation wavelengths. The FlowCentre™ workstation, a high performance multi-media platform incorporates Intel® Pentium® processing chip technology. The system features EXPO™ Software for instrument control, data acquisition and analysis in a Windows 95 environment.

MoFlo® Cytometers

Cytomation http://www.cytomation.com/index.htm has developed diverse systems for flow cytometry. The MoFlo® MLS Flow Cytometer is a high-performance cell sorting system for high speed sorting applications. In addition, Cytomation produces a MoFlo® BTA cytometer for bench top analysis, the MoFlo® BTS benchtop sorter and the MoFlo® SX cytometer for X/Y sperm sorting separation.

The MoFlo incorporates parallel processing electronics first developed at the Lawrence Livermore National Laboratory (LLNL) and was used successfully as a chromosome sorter for the Human Genome Project. It was designed first and foremost as a high-speed sorter and analyzer, offering the sort of performance needed for the quantification and/or isolation of rare cell populations with high recovery and purity at speeds in excess of 25,000 cells per second. Current performance specifications claimed by the manufacturer are a sorting speed of 50,000 cells per second, with purity and recovery greater than 99 percent.

Cytomation offers a number of accessories for the MoFlo MLS cell sorter. 4Way™ Sorting allows researchers to sort up to 4 subsets of cells at one time, using a combination of scatter and fluorescent parameters. This new mode of sorting allows sorting of more subsets faster, and with fewer cells lost to waste.

To ensure accurate cell sorting Cytomation offers Sortmaster™ an automated system that determines the droplet delay for each sort. The system monitors the droplet break-off during the run, ready to interrupt the sort and determine a new droplet delay if an instability occurs. If a stable break-off point is found, the sort resumes. If not, the sort will be aborted and the operator notified by an audible alarm and/or pager.

CyCLONE® is an automated high-speed cloning device that sorts into trays of 96, 384 or 1536-well configuration. In addition, the software that controls the CyCLONE unit can be configured for any user specified pattern. A 96-well tray can be filled with single cells in under 50 seconds.

Specifications and Characteristics
Purity, recovery and yield of sorted cells are important parameters that researchers need to be concerned with. The ultimate goal is to obtain the highest values for each of these parameters. In the real world, there is usually a trade-off between these criteria, especially when increasing the sort speed. The MoFlo MLS is capable of generating outstanding purity results such as these at 25,000 events/second. MoFlo's electronics technology increased speed of sorting without compromising recovery and yield, making the customary problem of turning up the sort speed only to degrade recovery and yield a problem of the past.

One measure of the precision of a cell sorter's performance is the coefficient of variation, simply abbreviated as CV. This is the standard deviation of the mean over the mean. It is a function of how well the optical elements are aligned with the sample stream. When multiple lasers are focused through a single optical element, only one beam can traverse the optical center of the lens. The other beam geometries may be compromised. This situation is best seen by noting the larger CVs generated by the off-axis beam vs. the on-axis beam. The MoFlo eliminates this situation by using separate optical elements for each beam. All beams are spatially separated and optically optimized.

Non-commercial produced flow cytometers

The "HiReCS" system Flow Cytometry/Cell Sorting at the University of Texas
The University of Texas has developed diverse technologies in the field of cytometry allowing them to construct their own flow cytometers. The University of Texas facility consists of a home-built, High-Resolution Cell Sorter (the "HiReCS" system) equipped with two lasers. A dual argon-ion laser/dye laser system (a 5 watt UV-enhanced, water-cooled Spectra/Physics Model 2025 argon-ion laser and a 300mW air-cooled argon-ion laser (Ion Laser Technology Model 5500AWC) and a Spectra/Physics Model 350 dye head unit which is optically split to provide up to three excitation laser beams at two spatial locations. This configuration provides up to 6-color fluorescence analysis capabilities. Complex 3- and 4-color fluorescence experiments are routinely performed.

Unlike most commercial systems, a home-built computer-controlled 3-color compensation system allows for simultaneous correction of three overlapping fluorescent signals. A single-cell sorting module allows for sterile cloning of individual cells and allows sorting of single cells directly into tubes for subsequent PCR enzymatic amplification of DNA or RNA. A number of other home-built, patented devices are incorporated into the system. These include a high-resolution time-of-flight measurement system capable of sizing cells and subcellular organelles as small as 0.3 microns in length or diameter; a multiparameter high-speed "rare-event" analysis system ("HISPEED") capable of analyzing cells at rates in excess of 100,000 cells per second. With this device, analysis of rare cell subpopulations as small as 0.0001 percent with sample rates of more than 10^9 cells per hour can be achieved.

A home-built sophisticated data acquisition system with Windows™ graphical user interface runs on a 66 MHz 80486 computer and allows for acquisition of 8-parameter, 12-bit listmode data first into RAM, then onto a 200 megabyte disk at rates in excess of 10,000 cells per second. The advanced and unique HiReCS also permits visualization, analysis, and sorting of cells from very complex multi-color fluorescence experiments. Correlation between multiple flow cytometric parameters can be explored through the use of multidimensional principal component/biplot analyses and associated real-time sorting. These analysis and sorting methods can also be used to reduce false-positives. A unique expert system has been developed providing "flexible sorting" strategies to optimize sample yield or purity. Instrumentation is continually designed or updated in response to particular biological applications.

Partec™ Flow Cytometers (CyFlow® and PAS®)

Partec (http://www.partec.de) has develop the CyFlow®, a compact flow cytometer for diverse flow cytometry analysis including the capability to determine absolute volumetric counting. The CyFlow allows to analyze FSC, SSC and up to three fluorescent signals. Since the FloMax® software is PC based, the data acquisition, analysis, and real-time display is performed with any type of PC or laptop within a windows platform. The software allows postacquisition compensation, a very important feature during the analysis.

PAS® is a more sophisticated Compact Flow Cytometer from Partec and it is designed for Multiparameter Flow Cytometry analysis for applications in Immunology,, Microbiology, Biology, Pathology, Hematology including Lymphocyte Subset and Cell biology Analysis

Equipped with an Argon ion laser (488 nm) and an HBO lamp for UV excitation it can determine up to 8 parameters: FSC, SSC, FL1 to FL6 (FITC, PI, DAPI, SR, TR, PE-CY5, AMCA, APC, and many more colors and combinations)

It can detected less than 100 FITC molecules, and determine the true volumetric absolute counting that is very useful for assess CD 34+ , CD4/CD8 and other immunolabelled cells without need for reference particles.

The Partec flow cytometers are very robust and compact as well as easy to handle and less demanding in location space, maintenance and service. In addition, the fact that some Partec flow cytometers like the CyFlow® runs on 12 V DC power (e.g. on a car battery) can be important for users in remote places or in field research and for developing countries.

3

Flow Cytometric Data Analysis. CellQuest Software

This chapter sequentially describes how flow cytometric data is analyzed and presented graphically in the form of contour plots, dot plots and histograms. Moreover, this chapter about flow cytometric data analysis shows some features of CellQuest software. CellQuest is the software developed by BD with capabilities for acquiring and analyzing samples. It is designed in a user friendly way and permits the performance of off-line analysis.

Gating
The goal in analyzing flow cytometric data is to define the percentage of cells expressing the marker(s) under investigation. Data analysis usually defines these cells by comparing the analyzed samples with appropriate negative controls. The first step in this process is to specify or define the cells of interest. This step is defined in flow terminology as gating. Gating traditionally is performed based on the physical characteristics of cells as determined by forward and side scatter signals. In Figures 5a and 5b, the upper left panel represent a dot plot graphic of a dendritic cell line that shows forward and side scatter and a gate (R1) is set up in the dendritic cell region of the sample. The cells were stained with monoclonal antibody conjugated to FITC. In Figure 6-9 the upper left panels represent a dot plot graphic that shows the forward and side scatter of cat's PBMC and a gate (R1) is set up in the lymphocyte region of the sample. However, the current and more elaborate gating strategy utilizes a combination of one fluorescence parameter along with a scatter parameter. In situations where it is important to construct an accurate lymphocyte gate to exclude the possibility of contaminating cells, *i.e.*, monocytes, side scatter is combined with an anti-leukocyte conjugated monoclonal antibody. Since reactivity against the leukocyte common antigen (CD45) has been shown to identify bone marrow derived cells, CD45 versus side scatter can be used to set gates around CD45+ lymphocytes in peripheral blood for the enumeration of CD4/CD8 lymphocyte subsets. The CD45 versus

side scatter gating strategy has been used to identify blast populations in bone marrow for leukemia phenotyping. The gates are drawn in the dot plot either as a rectangle, an ellipse, or a multi-sided polygon encircling the population of interest. Cells within the defined gating boundaries can be quantified and the expression of the cellular marker(s) in question can be depicted in histograms (Figure 5a and 5b), contour plots (lower right panel of Figures 6-9), density plots (upper right panel of Figures 6-9) and dot plots (lower left panel of Figures 6-9)

Histograms and the setting of markers
Most of the applications involving samples labeled with fluorochrome(s) yield data that are best presented using a single-parameter histogram with the x-axis representing fluorescence intensity and the y-axis representing the number of events per channel (of a specific fluorescence intensity) along the x-axis (Figure 5a upper and lower right panels). The majority of histograms representing immunofluorescence data use a logarithmic scale from 10^0 to 10^4 (Figure 5a and b right panels). This form of graphic display is ideally suited for making a comparison between the negative (control) sample (Figure 5a, lower right panel) and the test sample (Figure 5a upper right panel). Thereafter, histograms displaying successive samples are enumerated and compared by one of several types of statistics like Kolmogorov-Smirnov statistics (Figure 5b, lower panels). In the histogram of the negative sample, a cursor (marker) is set at the upper end of the left-most, "negative" distribution (M), to include between 95% - 99% of all events. This results in approximately 1% to 5% of the negative control events appearing to the right of this cursor as being considered considered positive (M1 in Figure 5b upper right panel). Defining the upper and lower boundary with the cursors is referred to as the setting of markers. This type of representation and the statistics obtained is widely used for single color analysis of PBMC's or during reticulocyte analysis.

In situations where a single peak may or may not be gaussian in shape, an "analysis region" is drawn to encompass the peak. The statistic associated with this region is either a mean or median channel position which is directly proportional to the fluorescence intensity (MFI) of the population being quantified. In this approach, the MFI of the negative sample is compared to the MFI of the test sample (Figure 5b lower panels). This type of histogram representation and associated statistic is also used for single color analysis of PBMC's and platelet-associated Ig (PAIg). When the fluorescence distribution shows several peaks which is a characteristic of DNA content analysis, an additional type of representation is used. Since the peaks may represent different DNA ploidy, several analysis regions encompassing the populations are drawn. Each ploidy distribution may denote different stages of the cell cycle. In such cases, the mean channel position, is associated with another statistic, the coefficient of variation (CV), *i.e.*, the standard deviation divided by the mean, which is a common measure of precision of each peak.

Dot Plots

Dot plots are commonly used to correlate any two parameters of a given cell population by displaying them in a two-dimensional dot plot (scattergram). Moreover, dot plots can provide sharp discrimination between negative and positive cells by creating at least four possibilities so that there is no overlap between the cells that are single or double negative for each marker with the cells that are bright enough to be single or double positive for the cellular markers in question. This is usually the case with CD4/CD8 determinations where a dot plot is divided into quadrants and the results are expressed as percentages of cells appearing within specific quadrants compared to all cells represented in the entire plot. Dot plots are appropriate if a sufficient number of cells are available for analysis and if the fluorescence intensity associated with the cellular markers is bright enough to discriminate between positive and negative (Figure 6-9, lower left panels).

The dot plot is a bivariate representation where each dot represents a single event (cell or particle) exhibiting multiple characteristics measured and recorded by the cytometer. Statistical analysis may be applied to four populations of cells as delineated by a quadrant marker. The four quadrants formed by the marker are labeled: LL, UL, UR, LR to indicate its location:

LL (lower-left) represents cells which are double-negative for the parameters represented on both the x- and y-axis.

UR (upper-right) represents cells which are double-positive for the parameters represented on both the x- and y-axis.

UL (upper-left) represents cells which are only positive for the y-axis parameter, but negative for the x-axis parameter.

LR (lower-right) represents cells which are only positive for the x-axis parameter, but negative for the y-axis parameter.

Since dot plots allow the identification of positive or negative populations in a bivariate representation, it is mandatory to have appropriate samples to serve as negative controls in order to accurately set up the position of the quadrants

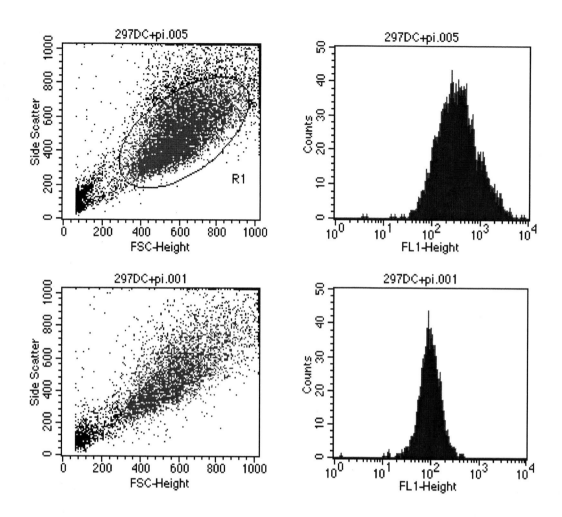

Figure 5a. Flow cytometric analysis of a human Dendritic cell line. Left panels, the FSC and SSC dot plots of a dendritic cell line. The cells were gated (R1) and the histogram of the gated dendritic cell lines are showed. In upper right panel is displayed the cells that were stained with anti DC monoclonal antibody conjugated to FITC. In lower panel is displayed the cells stained with isotype control.

and obtain the correct percentage of cells falling within each quadrant (Negative control shown in Figure 6). In addition, samples positive for only one of the markers should be prepared. Figure 7 and 8 shows the plots of PBMC's stained only with either CD4-FITC or CD8-PE from a feline blood sample (Single positive). The sample stained with both antibodies (CD4-FITC and CD8-PE) is shown in Figure 9.

Contour plots and Density plots

Data presented in the form of either contour plots or density plots is used

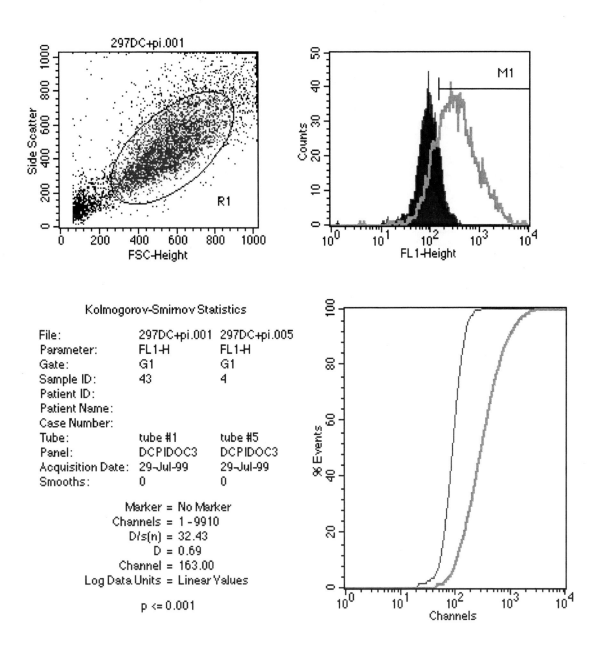

Figure 5b. Flow cytometric analysis of a human Dendritic cell line (histogram overlays and statistical analysis). Left panel, the FSC and SSC dot plots of a dendritic cell line. The cells were gated (R1). Upper right panel represents the histogram overlays of the gated dendritic cell lines stained with monoclonal antibody conjugated to FITC (open) or isotype control (solid). M1 is shown to define the negative and positive of the samples. Statistical analyses are shown in the lower panels.

when the emphasis of the analysis is on patterns of reactivity among diverse populations, rather than in the simple enumeration of percentages within each quadrant. Therefore, contour plots offer a qualitative assessment of the position

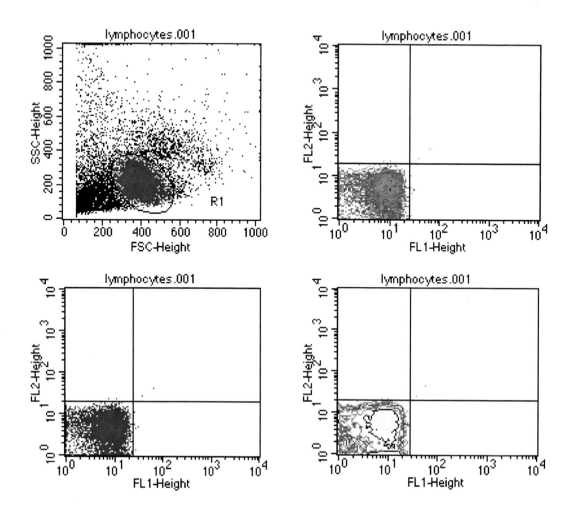

Figure 6. Flow cytometric analysis of feline PBMC. PBMC's stained with isotype control-FITC and isotype control-PE from a feline sample (dot plot representation is in lower left panel; density plot representation is in upper right panel; contour plot representation is in lower right panel).

of clusters within the plots not only in terms of being positive or negative, but also for expressing reactivity denoted by the relative fluorescence intensity (*i.e.*, weak, moderate, strong). Moreover, the coloring of each discernible contour level (ring) or density level (colour) distribution on the plot makes for easier visualization and interpretation (Figure 6-9, right panels). Two-parameter contour plots are frequently used for immunophenotyping of leukemia or lymphomas.

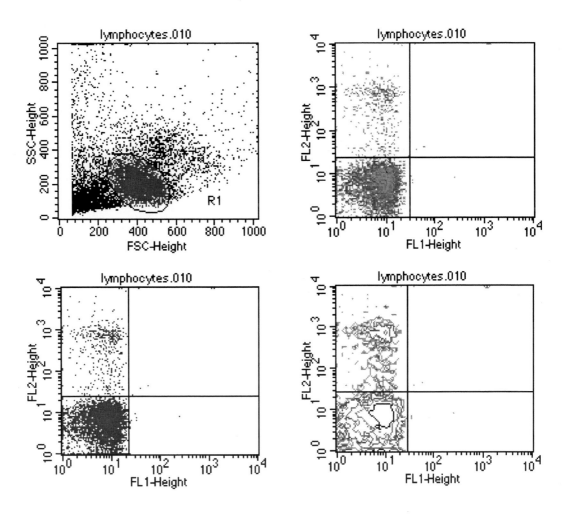

Figure 7. Flow cytometric analysis of feline PBMC. PBMC's stained with CD8-PE from a feline sample (dot plot representation is in lower left panel; density plot representation is in upper right panel; contour plot representation is in lower right panel).

Additional Software for data analysis-FlowJo™/WinMDi™

FlowJo™
FlowJo is a program designed to analyze data generated by any Flow Cytometer from any manufacturer. FlowJo has several analysis platforms that allow not only the performance of standard analyses such as gating and statistics, but also specialized analyses such as DNA/Cell cycle and kinetics. It has sophisticated tools for generating output material (graphical or tabular) with publication-quality output. In addition, it permits the export of any graphs,

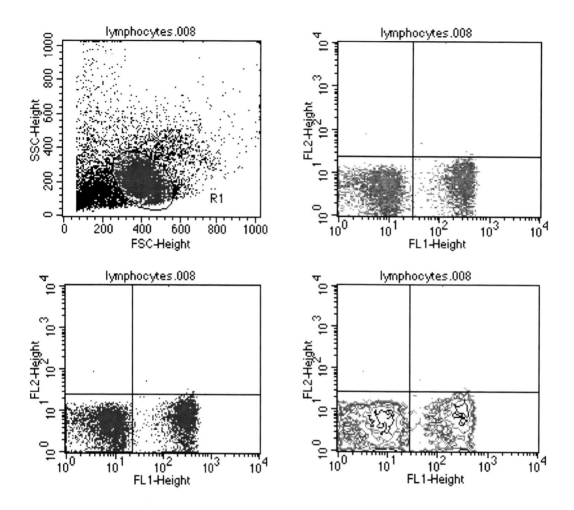

Figure 8. Flow cytometric analysis of feline PBMC. PBMC's stained with CD4-FITC from a feline sample (dot plot representation is in lower left panel; density plot representation is in upper right panel; contour plot representation is in lower right panel).

statistics, or other information into other programs for further analysis and presentation.

A detailed description of this software can be found at: http://www.treestar.com/flowjo. Some details of the program and of its features and benefits are listed below:

It performs complex analyses of up to thousands of samples containing millions of events. FlowJo is a Macintosh based software that makes it possible to do batching, simply and intuitively. Despite the Macintosh platform, it reads files

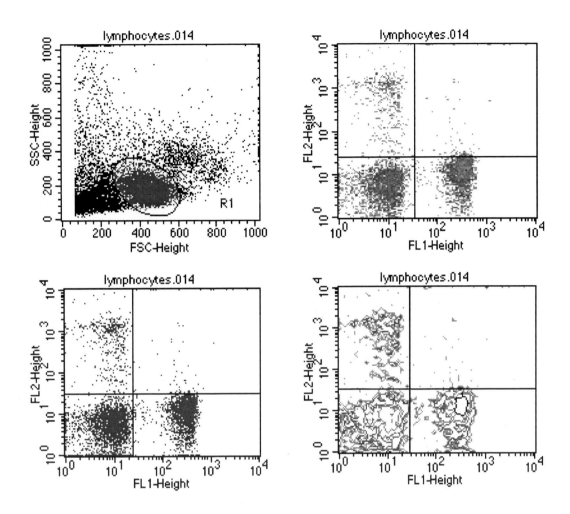

Figure 9. Flow cytometric analysis of feline PBMC. PBMC's stained with CD4-FITC and CD8-PE from a feline sample (dot plot representation is in lower left panel; density plot representation is in upper right panel; contour plot representation is in lower right panel).

from any flow cytometer, whether collected on a HP, PC or Mac.

FlowJo produces good publication-quality graphics, including a feature for presenting and publishing flow data on the WWW. Through a FlowJo's Groups structure, gates can be applied either to multiple samples or to a single one.

More specific features are described as follows:

i) Permits analysis, saving of gates, statistics, tables, layouts in a workspace document.

ii) Also does Dot plot, Contour plot, Density Plot, Pseudo coloring, Histograms.
iii) Software for compensation postacquisition.
iv) Enables multi-color, multi-laser data analysis.
v) Cell cycle analysis including: analysis of DNA, Cell cycle data.
vi) Advanced kinetics analysis.
vii) Strong batching generation of graphs and statistics while still retaining flexibility and control.
viii) Single drag-and-drop operations even when performing complex batch analyses.
ix) No macros or programming required.
x) Gated data in either FCS or spreadsheet format, easy copying of graphs to presentation packages enabling rapid production of publication-quality output.

FlowJo's Groups structure is really the powerful feature of this program: when an operation like a gate or a statistical analysis is performed on a group, it performs the same operation on every sample belonging to that group. Thus, that gate or analysis will be automatically performed on all samples.

An analysis with FlowJo require the following steps:

i) Create a workspace and load the samples.
ii) Create a group(s) of samples.
iii) Perform a thorough analysis on a single sample. Choose between gates, statistics, etc.
iv) Copy the appropriate analyses to appropriate group(s).
v) Check if the gates within the samples in the group are appropriate. Give enough room to accommodate sample variation.
vi) Generate a graphical layout containing graph(s) from all samples.
vii) Generate a table, containing statistics from all samples.
viii) Save this workspace. Next time, you can apply this workspace to the new samples. Thus all gates and statistics, etc. of your previous analyses will be automatically applied to each sample. You generate only the new graphical layouts and the tables.

WinMDi
WinMDi is a software designed for PC systems. It is available from the internet free of charge and has the capability for designing graphics with good quality for publication. Figure 5 of the paper by Kamau *et al.* is an example of the good performance produced by this software.

References
1. Givan A. 1992. Flow Cytometry: First Principles. Wiley-Liss, New York.
2. Owens MA, Loken MR. 1995. Flow Cytometry: Principles for Clinical Laboratory Practice. Wiley-Liss, New York.
3. Ormerod, M. Data Analysis in Flow Cytometry – a Practical Approach. A CD-ROM.
4. Kamau S., Hurtado M., Müller-Doblies U., Grimm F., Nunez R. 2000. Flow cytometric assessment of allopurinol susceptibility in *Leishmania infantum* promastigotes. Cytometry. 40: 353-360.

4

DNA Analysis. DNA Measurement and Cell Cycle Analysis

Two of the most popular flow cytometric applications are the measurement of cellular DNA content and the analysis of the cell cycle. Therefore, diverse protocols for DNA measurement have been developed including Bivariate cytokeratin/DNA analysis, Bivariate BrdU/DNA analysis, and multiparameter flow cytometry measurement of cellular DNA content. These analyses have been paralleled with the development of commercial software for cell cycle analysis.

Flow cytometry measurement of cellular DNA content
The nuclear DNA content of a cell can be quantitatively measured at high speed by flow cytometry. Initially, a fluorescent dye that binds stoichiometrically to the DNA is added to a suspension of permeabilized single cells or nuclei. The principle is that the stained material has incorporated an amount of dye proportional to the amount of DNA. The stained material is then measured in the flow cytometer and the emitted fluorescent signal yields an electronic pulse with a height (amplitude) proportional to the total fluorescence emission from the cell. Thereafter, such fluorescence data are considered a measurement of the cellular DNA content. Samples should be analyzed at rates below 1000 cells per second in order to yield a good signal of discrimination between singlets or doublets.

Since the data obtained is not a direct measure of cellular DNA content, reference cells with various amounts of DNA should be included in order to identify the position of the cells with the normal diploid amount of DNA. Some of the common reference cells often used for DNA measurements are human leukocytes or red blood cells from chicken and trout (Figure 10).

Commonly DNA measurements are expressed as a DNA index of the ratio of sample DNA peak channel to reference DNA peak channel. A DNA index of 1.0 represents a normal diploid DNA content, while deviations in cellular DNA content values other than 1.0 indicate DNA aneuploidy.

The analysis of the cell cycle

In addition to determining the relative cellular DNA content, flow cytometry also enables the identification of the cell distribution during the various phases of the cell cycle. Four distinct phases could be recognized in a proliferating cell population: the G1-, S- (DNA synthesis phase), G2- and M-phase (mitosis). However, G2- and M-phase, which both have an identical DNA content could not be discriminated based on their differences in DNA content (Figure 10 and 11). Diverse software containing mathematical models that fit the DNA histogram of a singlet have been developed in order to calculate the percentages of cells occupying the different phases of the cell cycle.

Approach to discriminate doublets or cellular aggregates

A common feature of DNA analysis is the finding of doublets or cellular aggregates. A doublet is formed when two cells with a G1-phase DNA content are recorded by the flow cytometer as one event with a cellular DNA content similar to a G2/M-phase cell. If a sample contains many doublets, that could mistakenly increase the relative number of cells in the G2/M-phase of the cell cycle, yielding to an overestimation of G2/M population. In order to correct this error, modern flow cytometers are equipped with a Doublet Discrimination Module that selects single cells on the basis of pulse processed data. The emitted fluorescent light of the DNA dye (FL2) generates an electronic signal that can be recorded as high (FL2H) for the intensity of the staining as well as measured as pulse-area (FL2A) and pulse-width (FL2W) of the samples. By plotting the FL2W versus FL2A in a dot plot graph a discrimination of a G1 doublet from a G2/M single can be made. Since the FL2W increases with the diameter of the doublet particle while both the G1 doublet and the G2/M single produce a same FL2A signal is likely to discriminate the doublet from the single (Figure 10). Therefore, in the dot plot graph a gate (G0/G1/S/G2/M) is set around the single population. The histogram graph of this gated population shows the four distinct phases that could be recognized in a proliferating cell population (Figure 11, upper right panel): the G0/G1-, S- (DNA synthesis phase), G2- and M-phase (mitosis). The number of cells in each phase is shown in Figure 11. G0/G1 is in lower left panel; S phase is in lower middle panel and G2/M in lower right panel.

Software for cell cycle analysis

Diverse manufacturers such as Becton Dickinson have developed software (CellFit™) for cell cycle analysis. The diverse software provides several mathematical models for fitting the DNA histogram. However, a subtraction of the background is required in order to remove events due to debris and to get a better fit with the models. Before the actual calculation of the phase distributions, two regions that are marked at the left and at the right part of the histogram are examined. Then, the data are fitted into an exponential curve of the form $y=e^{(ax+b)}$, and then the portion of the histogram that includes the two regions is subtracted from this curve. CellFit uses this approach to subtract the background.

However when using modeling for cell cycle statistics, there is a lot of variability and a factor of error because minute variation in the sampling and preparative techniques of the cells can contribute to inaccurate estimates. Therefore, it is essential to be aware of that variability in the analysis and interpretation of DNA content histograms. Also, it is important to evaluate the advantages of each software package and their respective limitations before using one for the analysis and fitting in the cell cycle models.

Bivariate BrdU/DNA

Flow cytometric analysis of cell DNA content is widely used for the estimation of cell cycle phase distributions. However, this analysis does not provide cytodynamic information such as cycle traverse rates and phase transit times. These parameters can be obtained using autoradiographic techniques, in particular, by determining the fraction of labeled mitosis. By autoradiographic identification of labeled cells during various intervals after labeling, their transition through the subsequent mitotic divisions can be followed, and phase and cycle transit times can be calculated.

In the last few years, non autoautoradiographic methods for distinguishing DNA synthesizing cells have been developed. These use monoclonal antibodies to measure the incorporation of bromodeoxyuridine (BrdUrd) into cellular DNA. In an indirect immunocytochemical technique, cells, which have incorporated BrdUrd, can be labeled with a fluorochrome and simultaneously the DNA can be counterstained. Flow cytometry then allows the simultaneous measurement of incorporated BrdUrd as well as the DNA content on a single cell level. In this way the cohort of labeled cells can be followed through the cell cycle.

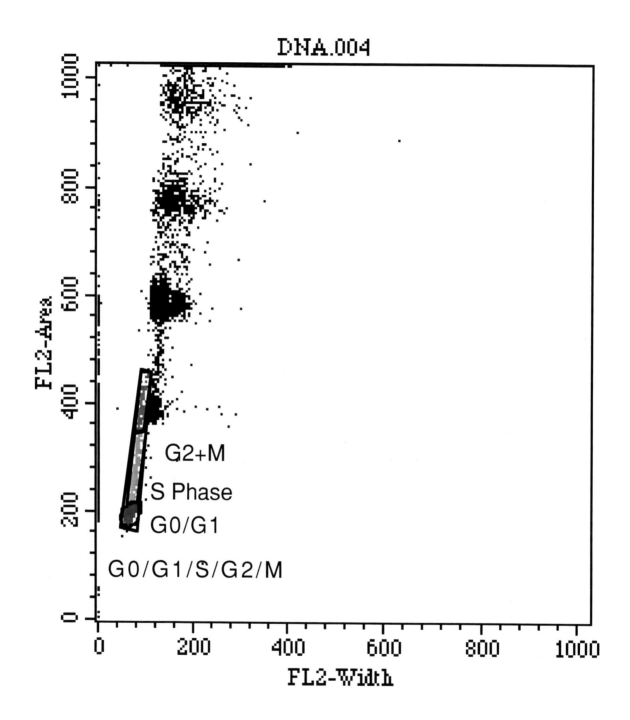

Figure 10. Dot plot of FL2W/FL2A for DNA content and doublet discrimination. An example of doublet discrimination in a dot plot of FL2W/FL2A. The G0/G1/S/G2/M population is gated and colored while the doublets remain uncolored.

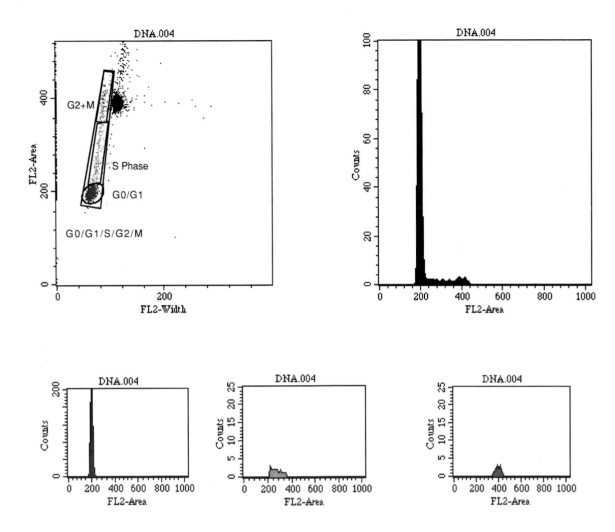

Figure 11. DNA and cell cycle analysis. An example of DNA and cell cycle analysis. Upper left panel shown the dot plot of FL2W/FL2A. The G0/G1/S/G2/M population is gated. Upper right panel shown the histogram of the gated G0/G1/S/G2/M cells: the G0/G1-, S- (DNA synthesis phase), G2- and M-phase (mitosis). The number of cells in each phase is shown in Figure 11 lower panels. G0/G1 is in lower left panel; S phase is in lower middle panel and G2/M in lower right panel.

DNA content analysis

In addition to analysis of cellular DNA content in lymphoreticular and hematopoietic neoplasm, flow cytometry has been used to characterize solid tumors. The most frequent tissues analyzed are a) biopsies from breast tumors, and b) tissue derived from products of conception. However, the value of DNA content analysis as a prognostic indicator in the case of tumors has been rather inconsistent and its use is being restricted to specific entities like breast cancer.

Staining is with a nucleic acid dye like propidium iodide (PI) and is similar to that described so far for other DNA assays. However, there are several circumstances in the staining procedures and analysis that require careful attention in order to avoid false interpretations. For example, since RNA would interfere in the staining, the solution should also contain RNAase. Also the number of cells acquired are critical in order to ensure that adequate cell numbers are collected. In addition, non-specific low-level staining must be excluded and only strongly stained cells should be collected in order to perform an accurate modeling of the data. Also as a part of the analysis, it is important to exclude debris and aggregates, as they will interfere with the measurements. Moreover, in some other circumstances, the G0/G1 peak of one ploidy distribution may be localized in the same area as the S-phase component of another distribution. In those circumstances when there is more than one ploidy distribution it is almost impossible by drawing a few analysis regions to delineate the cell cycle compartments. Therefore, the analysis and quantification of cell cycle compartments is a pretty complex task that requires the use of software for modeling.

An example of a typical analysis is as follows:
Stained sample is collected on the cytometer and displayed for qualitative assessment. The gating strategy varies depending of the type of software but always includes a step to exclude aggregates. Thereafter, FL2W and FL2A are plotted as either dot plot or contour plot. A gate is set up in the area of cells with 2N DNA content. Moreover, there still exists the possibility that one of the cells containing 4N DNA content could be located inside the single gate, or that doublets with 2N DNA content will be located outside the gate. A one-parameter histogram of FL2-A (PI fluorescence) is drawn from the cells with 2N DNA content. Usually two major peaks are observed; one peak is labeled as diploid and included in the region R1 that is colored in red and the other peak is labeled as aneuploid and included in the region R2 that is colored in green. In the further analysis, R1 is always depicted in red and R2 is depicted in green and any cells not associated with these two regions appear white. A SSC/FL2 dot plot is drawn which shows that the debris was stained neither in red nor in green but in white, facilitating its subtraction out of the analysis.

The report shows the separation and quantification of the two ploidy distributions: diploid and aneuploid, as well as the calculation of the cell percentages in each cycle compartment. Also it shows the CV's for the G0/G1 peak of each distribution, and a measurement of the DNA Index (DI), which is the aneuploid/euploid DNA content ratio.

References

1. Ross JS. 1996. DNA Content analysis. In: DNA Ploidy and Cell Cycle Analysis in Pathology. Igaku-Shoin, NY.
2. Any basic chapter on cell growth and division. Chapter 13 in Molecular Biology of the Cell by Alberts *et al.* (Garland Press) is a good one, but most cell biology text books would have an appropriate section.
3. DNA analysis and flow cytometry. Chapter 4 in Flow Cytometry: A Practical Approach edited by MG Ormerod (IRL Press).
4. DNA Modeling. Chapter 7 in Flow Cytometry: First Principles by AL Givan (Wiley-Liss).
5. Bagwell CB. 1993. Theoretical aspects of data analysis. In: Clinical Flow Cytometry Principles and Application. Ed Bauer KD, Duque RE and Shankey TV. Pages 41-61.
6. Rabinovitch PS. 1993. DNA Analysis Guidelines. In: Practical Considerations for DNA Content and Cell Cycle Analysis. Ed Bauer KD, Duque RE and Shankey TV. Pages 117-142.
7. Bauer KD, *et al.* 1993. Cell cycle antigens. In: Guidelines for Implementation of Clinical DNA Cytometry. Vol 14, No. 5, Pages 472-477.
8. Bauer KB and Jacobberger JW. 1994. Analysis of intracellular proteins. In: Methods in Cell Biology: Flow Cytometry Vol 41, 2nd edition, eds. Z. Darzynkiewicz, H. A. Crissman, J. P. Robinson, Academic Press, Inc., NY, pp351-376.
9. Srivastava P, Sladek TL, Goodman MN and Jacobberger JW. 1992. Streptavidin-based quantitative staining of intracellular antigens for flow cytometric analysis. Cytometry. 13: 711-720.
10. Sladek TL and Jacobberger JW. 1992. Dependence of SV40 large T-antigen cell cycle regulation on T-antigen expression levels. Oncogene 7: 1305-1313.
11. Sladek TL and Jacobberger JW. 1993. Flow cytometric titration of retroviral expression vectors: Comparison of methods for analysis of immunofluorescence histograms derived from cells expressing low antigen levels. Cytometry. 14: 23-31.
12. Sladek TL and Jacobberger JW. 1998. Cell cycle analysis of retroviral vector gene expression during early infection. Cytometry. 31: 235-241.
13. Sramkoski RM, Wormsley SW, Bolton WE, Crumpler DC, Jacobberger JW. 1999. Simultaneous detection of cyclin B1, p105, and DNA content provides complete cell cycle phase fraction analysis of cells that endoreduplicate. Cytometry. 35:274-283.
14. Jacobberger JW, Sramkoski RM, Wormsley SB, Bolton WE. 1999. Estimation of cell cycle-related gene expression in G1 and G2 phases from

immunofluorescence flow cytometry data. Cytometry. 35: 284-289.
15. Hedley DW, *et al*. 1999. Consensus review of the clinical utility of DNA cytometry in carcinoma of the breast. Cytometry. 14: 482-485.
16. Hedley DW. 1993. Breast Cancer. Ed Bauer KD, Duque RE and Shankey TV. Pages 247-261.

5

Molecular Cytometry

Molecular cytometry is a relatively new and versatile technique for addressing cellular and molecular biology questions. It achieves this by combining a number of molecular biology techniques in order to obtain quantitative molecular measurements of single cells. In addition, it permits the analysis of cell-to-cell variations in the molecular parameters being studied.

Some Molecular Cytometry techniques frequently used in flow cytometry include:
• Flow cytometry/cell sorting for cell micromanipulation.
• FISH (fluorescence *in situ* hybridization) in cytometry.
• PCR (polymerase chain reaction) and TaqMan.
• Image/Confocal Microscopy.

Flow cytometry/cell sorting for cell micromanipulation
Single-cell manipulation and microinjections are routinely performed with a Leitz inverted phase-fluorescence microscope equipped with a micro-manipulation/micro-injection system for: (a) manipulation of sorted cells on an individual basis for molecular characterization of gene products; (b) microinjection of probes for chromosome micro-dissection; (c) microinjection for gene transfer studies. Such an approach permits single-cell manipulation to characterize single sorted cells at a molecular level, or even identify cell subpopulations within a cell mixture by using PCR techniques.

Single-cell gene transfer is a technology employed by diverse groups. Moreover, novel methods of microinjecting genes into single cells using an amplicon based system are being developed to transduce cells with external genes such as GFP or β-gal (Figure 12, panel A). The transduced cells can be accurately quantified by multiparametric flow cytometric approach (Figure 12, panel C).

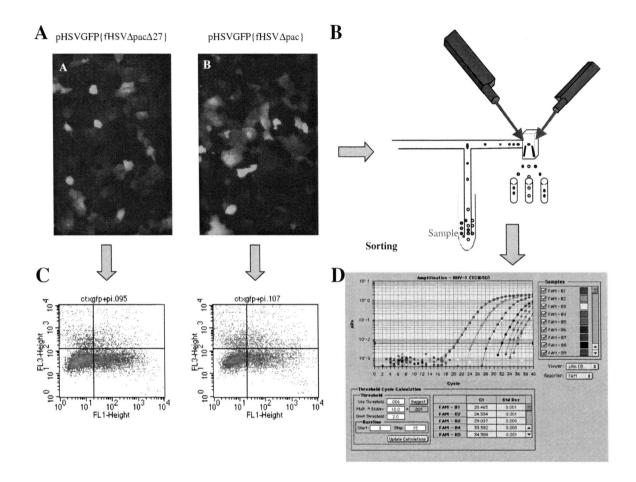

Figure 12. Molecular cytometry procedures. Panel A shows the transduction of cells with two different HSV-1 amplicon vectors. Cells were sorted and plated at diverse numbers as showing in panel B. The transduced cells were also evaluated by multiparametric flow cytometry as is shown in panel C. cDNA were prepared from the sorted cells (panel B) and real time PCR-TaqMan were performed (panel D).

FISH ("fluorescence *in situ* hybridization") in cytometry

There are numerous applications of FISH in cytometry. Initial work is performed by construction of microdissected chromosome probes utilizing an inverted phase/fluorescence microscope, a micromanipulator and micropipettes. The microdissected material is then amplified by PCR and labeled with reporter molecules suitable for subsequent painting of metaphase or interface chromosomes.

Another application uses chromosome "painting" in order to determine chromosome copy number. Also, FISH is used for detecting chromosome aberrations and/or damage as well as the location of sex-mismatched or species-mismatched donor transplanted cells in the tissues of a graft recipient. Two-color FISH is performed on metaphase chromosomes and interphase nuclei. Either flow cytometry or confocal microscopy can then examine the interphase nuclei.

PCR and real time PCR in molecular cytometry

The polymerase chain reaction (PCR) is an enzymatic amplification of specific DNA or mRNA sequences obtained from sorted cells. Most PCR assays are performed on either a programmable thermal cycler for sorted cells or on a thermocycler adapted for *in situ* PCR. *In situ* PCR has being developed as a way of amplifying small target sequences within individual cells for subsequent analysis by flow and image cytometry. PCR is also used for the construction of new DNA probes that will be used subsequently in FISH analysis. For example the PCR amplification of human DNA inserts in vectors and micro-dissected chromosomes in order to construct subregion-specific probes for chromosome painting.

Recently, real time PCR technique has been developed by coupling TaqMan™ technology to a sequence detection system (PE Applied Biosystems™), allowing the quantification of target sequences in real time. This development prompted the team up of a cytometry approach such as cell sorting with TaqMan yielding a highly accurate quantitative measurement of target sequences (Nunez R., Vögtling A, unpublished data) (Figure 12, panel B and D).

Image and confocal microscopy analysis

The new confocal systems provide high-resolution, 3-D color displays and full animation capability of confocal images. This equipment is complementary with a Zeiss Axiovert with CCD camara that provides routine digital microscopy. By using Adobe PhotoShop software, it is possible to adapt the image/confocal microscopy data to read the image files for analysis and preparation of material for publication and presentations (Figure 12, panel A).

References

1. Patterson BK, Goolsby CL, Hodara V *et al.* 1995. Detection of decreased CD4 expression in CD4 positive HIV-1 DNA positive cells by immunophenotyping and fluorescence *in situ* polymerase chain reaction. J. Virol. 69: 4316-4322.

2. Patterson BK, Mosiman VL, Cantarero L, Goolsby CL. 1998. Detection of persistently productive infection of monocytes in the peripheral blood of HIV positive patients using a flow cytometry based FISH assay. Cytometry. 31: 265-274.
3. Trask, B, van den Engh G, Pinkel D *et al.* 1998. Fluorescence *in situ* hybridization to interphase cell nuclei in suspension allows flow cytometric analysis and bicolor fluorescent *in situ* hybridization to lymphocyte interphase nuclei. Human Genet. 78: 251-259.
4. Nunez R, Ackermann M., Saeki Y., Chiocca A., Cornel F. 2000. Flow cytometric assessment of transduction efficiency and cytotoxicity of herpes simplex virus type 1 (HSV-1)-based amplicon vectors. Cytometry. In press.
5. Gray J.W. and L. S. Cram. 1990. Flow karyotyping and chromosome sorting. In: Flow Cytometry and Sorting, 2nd Edition (Mendelsohn, M., Lindmo, T.,and Melamed, M., ed.) Wiley-Liss Inc. New York, pp 503-529.
6. Weier, H.-U.G., D. Polikoff, J.J. Fawcett, K.M. Greulich, K.-H. Lee,L.S. Cram, V.M. Chapman, and J.W. Gray. 1994. Generation of five high-complexity painting probe libraries from flow-sorted mouse chromosomes. Genomics. 21: 641-644.

6

Surface Staining and Immunophenotyping Using Multicolor Analysis

Most benchtop analyzers are equipped with multiple detectors for performing multicolor analysis. Specimens can be stained with one, two, three and, with some instruments, four fluorochromes to perform multi-color analysis with a single tube (Figure 13). The larger and more expensive cell sorters can routinely perform five color analyses with the possibility of adding additional detectors. For specimens of small sample volume, multicolor analysis offers the researcher the possibility of detecting multiple cell surface markers at one time.

Immunophenotyping of Leukemia and Lymphoma samples
Currently, leukemia and lymphoma specimen are customarily submitted directly from the clinician for research, cell typing or for confirmation of a diagnosis. Specimens originating from peripheral blood, bone marrow, solid tissue, effusions (pleural or peritoneal centeses) and CSF for immunophenotyping can be processed. There are specific panels for markers of leukemia and for lymphoma; each type of specimen (*i.e.*, peripheral blood or bone marrow) presents different indications for phenotyping as well as the procedure for processing (1-9).

Submission of peripheral blood
The two main indications are:
i) the lymphoma panel is used for suspected chronic lymphocytic leukemia (CLL), and ii) the leukemia panel is used if circulating blast cells are present.

Submission of Bone marrow
The requisition order will usually specify "Research and typing" or "Confirmation" of leukemia or lymphoma. In some instances, the requisition

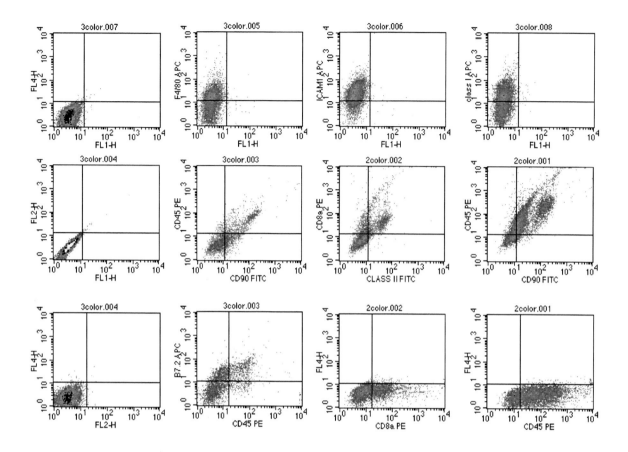

Figure 13. Multicolor analysis. On the left, three panels represent the staining of cells with isotype control negative antibodies. The two sets of panels in the center and the panel on the right show representative samples of cells staining with two or three colors.

will be non-specific in terms of classification. In these cases, the specimen(s) should not be processed until the history and diagnosis are given by the treatment team. In cases of studies with follow-up or treatment evaluation, it is helpful to check previous results in order to set up appropriate data analysis gates.

Submission of solid tissue
The sample is obtained during a surgical procedure; unless otherwise indicated, a lymphoma is assumed and a corresponding panel should be used.

Submission of effusions
Samples derived from Pleural or peritoneal centeses are normally submitted as part of a multi-section requisition for Cytology and/or Hematology

evaluation. All submissions should be processed and a lymphoma panel is used unless otherwise specified.

Submission of CSF
Specimens of CSF have special requirements since they frequently lack sufficient numbers of cells to complete an entire panel. Therefore, only few markers should be run. The decision to determine which markers are the most appropriate to use is based on the cellularity of the specimen and the presumptive or "rule-out" diagnosis. The CSF specimen submitted is pelleted and resuspended to between 500 µl and 1ml, depending on the cellularity of the specimen. At least 3-color staining is performed with CD4 and CD45 being key markers.

As general rules for leukemia/lymphoma phenotyping in fluids such as effusions and CSF, it is important to determine the CD45 and CD4 populations. Thus, a 3 or 4-color staining procedure that includes CD45 and CD4 is frequently used. Since the relevant cellular compartment is mainly peripheral blood, the fluorescence intensity of the CD45 and CD4 markers is sufficiently bright for easy separation of positive cells from negative cells. In addition, this approach makes it fairly simple to compare during the analyses the sample results with CD45 and CD4 populations with blood obtained from "normal" subjects which provide the reference ranges.

General consideration for leukemia/lymphoma phenotyping
i) Antibody panels are fairly standardized and there are general guidelines in this regard, however, there is quite a wide variation in the way that antibodies are combined into "panels" and there is currently not a consensus for constructing these panels.

ii) While it is fairly common to examine peripheral blood, some samples are obtained from other substrates, therefore the concept of peripheral blood reference ranges is sometimes not appropriate since similar tissues (bone marrow, lymph nodes, pleural fluid, peritoneal fluid, or CSF) are not readily available from "normal" individuals.

iii) The separation of negative from positive is not easy as most of the important markers in leukemia and lymphoma phenotyping are either present at low density or not present at all in circulating cells.

iv) The designing of panels that combine antibodies to reveal diagnostic relationships for given disease entities are in the initial stages and there is need for further investigation. A typical combination panel for bone marrow staining from a patient with suspected CLL is the combination of CD5-FITC/CD19-PE/CD45-PE-Cy5/CD23-APC. Normally, after acquisition, a contour plot of the CD45/Side Scatter (SSC) permits the defining of the "lymphocyte gate". Moreover, in cases of CLL there is a preponderance of cells in the so-called mature lymphocyte region. Thus, a gate is established around that region. Thereafter, three plots are used to display combinations of the other antibodies *i.e.*, CD19/CD5, CD23/CD19 and CD23/CD45.

In CLL, all CD19-positive cells are also positive for CD5. The CD19-positive cells have a continuum of reactivity with CD23 but some CD19+ cells are unexpectedly negative for CD23 (CD19+/CD23-). Such "pattern" of reactivity is very characteristic in CLL. Moreover, most of the immunophenotyping reports of hemo-lymphatic cancers are moving away from quantitative reporting of isolated percentages of positive cells toward a more qualitative one where patterns of reactivity are also examined, with marker results being described in terms as positive or negative and the staining intensity as strong or weak. Also, attempts are made to determine if the positive markers resemble any of the characteristic patterns associated with well-described malignancies as well as to estimate the stage of maturation of these cells. Additionally, flow cytometry results should never stand alone as diagnostic entities in these disease states, rather, the results must be correlated with morphologic assessment and clinical evaluation (1-9).

Other applications of surface analysis and immunophenotyping

Reticulocyte analysis
Reticulocytes are currently measured in a flow cytometer using thiazole orange, a dye that binds to nucleic acid and stains the immature red cells. However, thiazol also stains leukocytes and platelets. The advantage with flow is obvious since the time for analysis of a sample is short (only a few seconds). Also there is a more accurate and precise count (more than 50,000 red cells could be measured in comparison to 1000 by microscopy). Last but not the least important feature of the flow measurement is the possibility of determining the relative maturity of the reticulocyte population by measuring the relative fluorescence intensity and generating a value called the immature reticulocyte fraction (IRF).

The basis for the reticulocyte flow assay is the staining of whole blood with thiazole orange; only cells that contain nucleic acids incorporate the dye and develop green fluorescence. However, the signal emanating from reticulocytes can be distinguished from that of the WBC and platelets since the reticulocytes are cells with different FSC and SSC and in addition have a much dimmer fluorescence intensity due to lower levels of nucleic acid.

It is essential to analyze two tubes of each sample. The first one containing thiazole orange, and the second one (negative control for background fluorescence) containing only buffer. The negative control tube will help determine whether there is excessive autofluorescence due to inclusions bodies (HJ bodies) or to the presence of drugs that increase autofluorescence.

Results are reported as both a percentage of reticulocytes (RTC) and as absolute number of reticulocytes (RTA). This is possible if an analysis region is set up that separates the cells staining with thiazole orange, and expresses this value as a percentage of all RBC that have been counted, as well as calculating the absolute reticulocyte number. There is some drawbacks to this test due to the level of background fluorescence of the dye and the practicability of the test. When the background fluorescence is too high the distinction between positive and negative is obscured.

Thiazole orange stains and contaminates the chamber and the tubing of the cytometer and necessitates the thorough washing of the instrument before any other applications can be run. Therefore reticulocytes from a batch of samples are usually measured just once a day, creating some restrictions for practicability. However the measurement itself is no more complex than a CBC. Moreover, reticulocyte enumeration is a test that is included in the analysis package in the Hematology cell counters.

Platelelet-associated Immunoglobuling (PaIg)

Flow cytometry can help to discriminate the changes in the platelet morphology due to pathological diseases. Figure 14, shows the presence of diverse morphological changes (detected in the FSC) and granularity (detected by SSC) of platelets in animals affected with thrombocytopenia. In addition, measurements of platelelet-associated Ig is sought in patients suffering autoimmune diseases such as lupus erithematous or immune-mediated thrombocytopenia (IMT).

Flow cytometry can detect either total IgG (including IgG contained in alpha-granules), or surface IgG (less than 1% of the total PaIg in normal subjects). Both fractions are ussually increased in patients with IMT. However, total IgG is also increased in patients with non-immune thrombocytopenia (NIT).

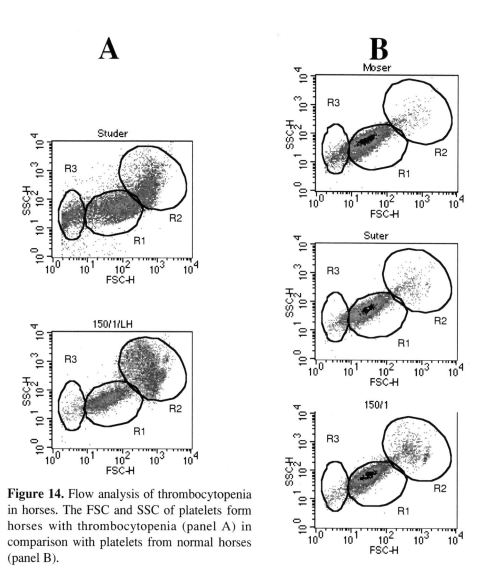

Figure 14. Flow analysis of thrombocytopenia in horses. The FSC and SSC of platelets form horses with thrombocytopenia (panel A) in comparison with platelets from normal horses (panel B).

Unfortunately, the flow measurement of platelelet-associated Ig only measures total Ig that may not be specific for immune-mediated disease. A test specific for immune-mediated disease, specifically in the context of drug-induced autoimmune thrombocytopenia, is not available through flow cytometry.

The current protocol is a two color staining of fixed platelets. In brief, platelet-rich plasma is extracted from EDTA-anticoagulated peripheral blood. Samples are centrifuged, and the platelets washed and fixed. Thereafter, a two-color staining is used. CD61-FITC provides the platelet-specific marker and allows the differentiation of platelets from other elements such as leukocytes, RBC, and sub-platelet particles. The second color used is a biotinylated goat anti-human Ig, followed by streptavidin-PE, and then the binding of immunoglobulin can be measured. During the cytometric data acquisition the CD61+ cells (platelets) are gated and the mean fluorescence intensity (MFI) of the PE signal of the gated cells is determined. Such MFI represents a relative measure of the bound Ig. This MFI data can be reported as raw data or a more sophisticated analysis can be performed using a standard curve, obtained after measurement of the MFI for PE of a mixture of PE-fluorescent beads with varying amounts of fluorescence (10).

Thus, in this approach the results are actually expressed as PE-fluorescence equivalents, rather than the raw MFI values. A disadvantage of this approach is that platelets from normal samples should be run in each assay since these platelets also have platelelet-associated Ig. However, after setting up and calculating reference ranges using samples of normal platelets, it is possible to differentiate negative from positive samples, and even based on the level of the measured fluorescence to divide the positive ones into strong, moderate and weak.

CD4 T cell enumeration and immunophenotyping for diagnosis of immunodeficiency or treatment follow up
Samples from patients with any suspected immunodeficiency condition, either innate or acquired, require a lymphocyte subset analysis. The test should include a measurement in terms of percentages and absolute counts of CD3+ (pan T cells), CD3+/4+ (T-helper), CD3+/8+ (T-cytotoxic/suppressor), CD16+/56+ (NK cell) and CD19+ (pan-B cells). Moreover, an increased proportion of requisitions for this measurement occur in the context of HIV-related conditions either during diagnosis or in monitoring disease course and therapy response in infected individuals.

CD4 eumeration is a routine test that refers to the quantification of CD4+ T-lymphocytes in peripheral blood. In this test the percentage and absolute CD4 count are the values of major interest. However these values should be provided in the context of a full cell blood count (CBC) including a report of absolute lymphocyte count with CD3 percentage and CD3 absolute count. In fact, multicolor analysis should be performed in this test.

The CD4 panel consists of 4 tubes. Each tube contains CD45 to gate the lymphocyte population. In addition each tube contains CD3 in order to provide a) quality control of reproducibility, b) an additional parameter to define a negative distribution during the setting of the quadrant regions, and c) to improve the T-cell specificity during CD4 enumeration since there are some populations of CD4+ cells that are not T lymphocytes and that are common contaminants in the lymphocyte gate. Of course each tube also contains CD4.

The analysis of these tubes generates results displayed as dot plots that represent cells that are within the lymphocyte gate (R1). The combination CD3/CD4 generate two CD4+ populations: one that is CD3- CD4+ and represents CD4 binding to monocyte and dendritic cells precursors. The other population is CD3+ CD4+, and represents the CD4+ T-lymphocytes (11).

Enumeration of CD34+ cells (Stem cells)
In order to assess the engraftment potential of bone marrow or peripheral blood stem cell transplants, flow cytometric methods of immunophenotyping and measurement of the cells bearing CD34 antigen are used. CD34+ cells are responsible for sustained multilineage ontogeny. However, the number of circulating CD34+ cells in peripheral blood from normal individuals is very low (in the range 0.01% to 0.1%). Therefore, diverse transplantation centers use a protocol that mobilize the cells containing the CD34 marker from the bone marrow by treatment of the patients with hematopoietic cytokines. Thereafter, the CD34+ cells are collected and enumerated by cytometry. Moreover, even with this mobilization protocol, the level rarely rises above 1.0%.

The enumeration and analysis of CD34+ cells is a cytometric challenge, since the cells are considered rare-events. A rare-event analysis has great implications in the way the assay is performed, because of background problems such as i) cellular autofluorescence, ii) Fc receptor binding of antibodies to CD34- cells such as monocytes, macrophages and dendritic cells, iii) non-specific binding to cellular debris. Also there are difficulties in setting up the markers for discrimination of negative/positive, as well as setting up the gates that include the CD34+ cells. In order to set up a standardized method for controlling all these variables Sutherland *et al.* recently published guidelines for CD34 analysis (12).

As the demand for CD34 enumeration has grown, several commercial sets of reagents and even a software called ProCOUNT™ has been developed. The major advantage of using a kit of reagents is in minimizing the variability by using standardized and titrated reagents. Also by using a customized software the instrument settings and gates are clearly defined. The software analysis utilizes automated algorithms in a strategy based on multiple, sequential gating that includes i) identification of nucleated cells with a nucleic acid dye, ii) determining the intensity for CD45 staining, and iii) side scatter signal. One kit even contains a bead standard for calculation of absolute counts of CD34+ cells. These types of kit are very popular in blood banks where they are used for measurement and preparation of peripheral blood stem cell products or for determining how many times a patient needs a pheresis. A typical analysis report displays the CD34 percentage, the ratio CD34+/ CD45+ cells, a traditional cell count and the pack volume, facillitating the determination of the number of CD34 cells collected per pack or pheresis and even per kilogram of body weight.

Immunophenotyping for T cells CD45RA/CD45RO
T cells from peripheral blood can be stained with either CD45RO or CD45RA. It seems that each marker stains a completely different T cell population. There is no overlapping between them and a lot of effort has been made in order to correlate such a strong separation of T cell subsets with a specific T cell function. CD45RA seems to identify naïve T cells while CD45RO identifies memory T cells.

Roederer *et al.* from the Department of Genetics at Stanford University have been working extensively on this topic in recent years (14), following seminal work by Louis Picker and colaborators. Recently Mitra *et al.* combined the CD45 with another set of T cell markers in order to gain more functional information and to further discriminate each subpopulation (13). Their results can be summarized as follows:

There is no further knowledge that can be gained by using both CD45RA and CD45RO in the same stain, because RA and RO stain T cell populations almost completely inversely. Selecting one or the other, and mixing it with other stains will provide further information about the population under study. For example adding CD62L, CD11a, or CD27, will yield more knowledge about the status of the T cell population under study. Among these, CD62L is the most likely of choice for fresh cells; while CD27 is preferable for frozen cells. All of them identify different memory subsets. However, they are almost identical in their capability to identify naïve T cells. In brief, naïve T cells has the following phenotype: CD45RA+, CD45RO-, CD62L+, CD11a-dull, CD11b-, CD27+, CD28+, CD57-, CDw60-, and CD95-. Interestingly, naïve T cells display lower amounts of CD28 than memory T cells. In addition, they express low (but detectable) levels of CD38. Of significance is the fact that naïve T cells are CD25-, HLA-DR-, CD69-, and CD71-. These are the so-called activation markers.

From this experimental data is drawn the following conclusions:

i) Almost all CD45RA+, CD62L+ T cells, are naïve. There is no naïve T cells in other phenotypes.

ii) Almost all CD45RA+, CD62L- T cells are memory T cells. This means they are neither naïve nor activated. These cells have already seen the antigen. Roederer *et al.* found that these cells carry the most restricted repertoire (14). This population contains the effector T cell at the most terminally-differentiated stage.

iii) CD45RA-, CD62L+ T cells are also memory cells. The principal population of IL-4 producing cells are within the CD4+, CD45RA-, CD62L+. This population is quite similar to naïve T cells in many functional aspects. Among these cells, the CD4+, CD45RA-, CD62L+, CD11a dull population displays the most Th2-like of all CD4 subsets. This subset has been found elevated in patients with Th2-diseases such as atopy or lepromatous leprosy, and diminished in Th1-disease such as tuberculoid leprosy.

CD62L produces a rather complex pattern of expression because within minutes of activation, CD62L is immediately clipped off by a surface protease upon PKC-activated signals. However, handling after freeze-thaw procedures also produce similar results. CD62L is then re-expressed at higher levels than resting after 24-48 hours. Later on, when the cells return to a basal status the pattern of expression depends on which subset the resting cells will eventually become.

iv) CD45RA-, CD62L- and CD11a+ cells are also memory T cells. In contrast to the CD4+, CD45RA-, CD62L+, CD11a dull population that produces IL-4, the CD11a-bright cells synthesized IFN-γ and is considered the most Th1-like of all subpopulations. In addition, this population is elevated in patients with Th1 diseases and diminished in Th2 diseases.

v) At the moment, there is no real data to fully understand the differentiation progression of these populations. There are some critical points still missing. For example, if there is back and forward progression between these populations, or if there is an ordered progression from one stage to the next with marker variation. Also it is not known if these subsets are generated by diverse differentiation steps from common progenitors.

vi) In addition, it is worth mentioning that there is a great degree of parallelism between the functions of any given phenotypic subset of CD4 and CD8 cells. For example, CD8+, CD45RA-, CD62L+ are the only CD8 IL4-producing cells as is the case in the CD4 cells. Also, the markers CD45RA+, CD62L- are found in CD8+ terminally differentiated toward a CTL function. This result suggests a more general principle that correlates surface markers with the function within the diverse CD4 and CD8 populations. Until recently, this set of markers has not been found in cells other than T cells. However, it has been recently found by Nunez *et al.* that CD45RO also stains a subpopulation of Dendritic cells (DC) and DC lines with full capability to present antigens to T cells, suggesting that CD45RO could identify mature DC, a finding that seems to parallel the data on T cells (15).

References

1. Stewart CC *et al.* 1997. U.S.-Canadian consensus recommendations on the immunophenotypic analysis of hematologic neoplasia by flow cytometry: Selection of antibody combinations. Cytometry 30: 231-235.
2. Jennings CD, Foon KA. 1997. Recent advances in flow cytometry: application to the diagnosis of hematologic malignancy. Blood. 90: 2863-2892.
3. Chan, J.K.C. *et al.* 1995. American Journal Of Clinical Pathology.
4. Stasi R. *et al.* 1995. Ann. Hematology. 71:13-27.
5. Viackus L, Ball E, Foon K. 1991. Critical Reviews In Oncology/Hematology. 11: 267-297.
6. Stewart C.C., S.J. Stewart. 1994. Cell preparation for the identification of leukocytes. In: Methods In Cell Biology, (Z. Darzynkiewicz, J. Robinson, H. Crissman, eds.) Academic Press, Inc. New York. 41: 39-60.
7. Stewart C. C., S.J. Stewart. 1994. Multiparameter analysis of leukocytes by flow cytometry. In: Methods in Cell Biology, (Z. Darzynkiewicz, J. Robinson, H. Crissman, eds.) Academic Press, Inc. New York, 41: 61-79.
8. Stewart C.C., S.J. Stewart. 1997. Titering antibodies. In: Current Protocols in Cytometry. (J.P. Robinson, Z. Darzynkiewicz, P. Dean. L. Dressler, P.Rabinovitch, C. Stewart, H. Tanke, L. Wheeless, eds.) J.Wiley & Sons, Inc., New York, 4.1.1 - 4.1.13.
9. Stewart C.C., S.J. Stewart. 1997. Immunophenotyping. In: Current Protocols in Cytometry, (J.P. Robinson, Z. Darzynkiewicz, P. Dean. L. Dressler, P.Rabinovitch, C. Stewart, H. Tanke, L. Wheeless, eds.) J.Wiley & Sons, Inc., New York, 6.2.1 - 6.2.15.
10. George JN. 1990. Platelet immunoglobulin G: its significance for the evaluation of thrombocytopenia and for understanding the origin of alpha-granule proteins. Blood. 76: 859-870.
11. CDC. 1997. Revised guidelines for performing CD4+ T-cell determinations

in persons with human immunodeficiency virus (HIV) infection. MMWR 46: 1-29.
12. Sutherland DR *et al*. 1996. The ISHAGE guidelines for CD34+ cell determination by flow cytometry. J. Hematotherapy. 5: 213-226.
13. Mitra DK. *et al*. 1999. Differential representations of memory T cell subsets are characteristic of polarized immunity in leprosy and atopic diseases. Int Immunol. 11: 1801-1810.
14. Roederer, M. *et al*. 1995. CD8 naïve T cell counts decrease progressively in HIV-infected adults. J. Clin. Invest. 95: 2061-2066.
15. Nunez R. *et al*. 1998. Characterization of two human dendritic cell-lines that express CD1a, take-up, process and present soluble antigens and induce MLR. Immunol. Lett. 61: 33-43.

7

Handling of Samples: Biosafety

The International Society for Analytical Cytology (ISAC) has created a committee that issues recommendations for biosafety in cytometry and image analysis procedures. Handling of biologic samples should be performed using latex gloves. Furthermore, a number of fluorescent dyes, *i.e.* rhodamine and propidium iodide are known carcinogens, therefore gloves must be worn when performing staining and sample analysis.

A disinfection agent such as sodium hypochlorite (household bleach) should be added to the waste tank of the FACSCalibur. When emptying the tank, gloves, lab coat and safety glasses or goggles must be worn to protect the researcher from splashes. The work area must be kept clean and should be disinfected with 70% ethanol or a dilute bleach solution. Spills must be immediately contained, disinfected and wiped up. After analysis, all surfaces that have the potential for contamination must be cleaned using 70% ethanol. This includes the FACSCalibur fluidic control panel, keyboard and mouse.

For the most recent information on biosafety guidelines and procedures, a biosafety page has been established within the ISAC web site. More information is available on-line at: http://www.isac-net.org/committees/biosafety/biosafety.html

Staining
The antigens are identified by staining with fluorescent probes such as conjugated antibodies, as in the case of cellular immunophenotyping, or with dyes in the case of reticulocyte enumeration and nucleic acid staining during DNA content analysis.

Every flow cytometric procedure commences with a sample that could be either cells or tissues, continues with staining and washing steps, proceeds through the setting up of the flow cytometer and the acquisition of flow data, and culminates with the analysis of the data that includes comparative statistics and the reporting of results. All these procedures require the setting up of appropriate controls and a rigorous quality control that can be obtained only by following reproducible protocols.

The cytometer itself should be set up and monitored routinely with a quality control program that involves the use of calibration standards for reference and control of the alignment. Therefore, every cytometric procedure represents a formal test for quality and reliability of the equipment, the operator and the institution.

8

Intracellular Antigens. Cytokines and the Study of Viral Antigens

Cytokines are proteins that play a critical role in the immune system. However, most of the detection systems for cytokine measurements use ELISA methods. Recently, it has been feasible to measure cytokines intracellularly by treatment of the cells with drugs that block the golgi and allow the cytokine accumulation making feasible detection by flow cytometric approaches. Diverse protocols have been developed for identification of intracellular cytokines. Moreover, Becton Dickinson has produced three documents that show the procedures for detection of intracellular cytokines in lymphocytes and in monocytes.

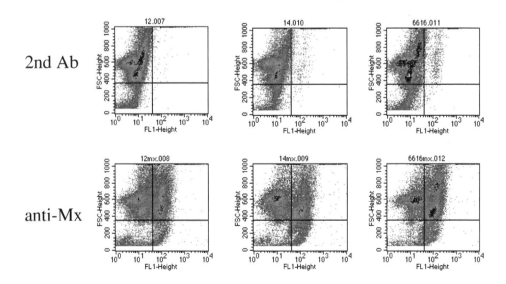

Figure 15. Intracellular expression of Mx in PBMC from Bovine cells. The intracellular expression of Mx on bovine PBMC stimulated with IFN alpha for 48 hours. Upper row panels show the samples stained with 2nd antibody only. Lower row panels show the samples stained with anti-Mx mAb and 2nd antibody.

Other intracellular antigens

The intracellular Mx protein is an interferon inducible protein. In addition, Mx is also induced during viral infection. Mx can also be detected by flow cytometry following intracellular staining. This procedure includes a step for permeabilization of the cell membrane. We have determined the intracellular expression of Mx in human PBMC, Bovine PBMC and Ovine PBMC (Nunez R, Metzler F. Unpublished data).

9

Flow Cytometric Assessment of Sperm Quality and Cell Cycle Analysis

In order to determine sperm quality, dual fluorescent staining of mammalian sperm coupled to a flow cytometric measurement has been developed (1). The flow cytometric estimation was performed by staining the spermatozoa with two fluorochroms stains, one was the membrane-permeating substrate carboxyfluorescein diacetate (CFDA) and the second was the relatively membrane-impermeant nuclear stain propidium iodide (PI). Mammal spermatozoa were evaluated and three distinct populations of spermatozoa were clearly identifiable in samples from bulls, boars, dogs, horses, mice, and men upon microscopic examination. A population of viable and motile spermatozoa retained the florescence chromophore throughout the cell, a second one retained the green florescence fluorophore mainly in the acrosome whilst the nuclei stained red with PI, and a third population consisting of degenerate spermatozoa showed only red fluorescent nuclei. These three populations were also identified and quantified in bovine spermatozoa by a dual parameter flow cytometric approach and this technique showed a strong correlation with other seminal quality measurements (1).

Further improvement in the flow cytometric estimates of sperm quality parameters such as viability, acrosomal integrity, and mitochondrial function, were achieved by a multicolour staining procedure (2). Bull spermatozoa viability was evaluated by measurement of propidium iodide (PI) staining, the acrosomal integrity was also measured by *Pisum sativum* agglutinin (PSA) binding to acrosomal contents and the mitochondrial function was assayed by rhodamine 123 (R123) fluorescence. It was found that by combining these staining regimes and measuring the spermatozoa by a flow cytometer, it was possible to correlate the three different features simultaneously on individual spermatozoa. In addition, thousands of cells per sample could be assayed in a relatively short period of time without extensive preparation (2).

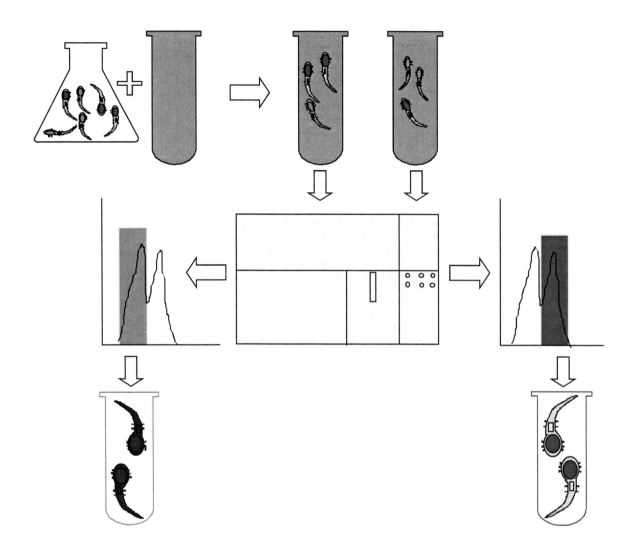

Figure 16. Flow cytometric approach to separate X and Y spermatozoa. The flow cytometric approach for assessing the proportion of X- and Y-spermatozoa in semen of either domestic animals or human subjets (upper and middle panels). The approach for sorting and enrichment of X-spermatozoa (lower right panel) or Y-spermatozoa (lower left panel).

An additional improvement in viability measurements is obtained by replacing the fluorogenic stain of the membrane-permeate substrate carboxyfluorescin diacetate (CFDA) with SYBR-14, a newly developed fluorescent nucleic acid stain (3). Since SYBR-14 absorbs at 488 nm and emits at 518 nm when bound to DNA, living sperm from six representative mammals could be assayed by dual staining with SYBR-14 and propidium iodide (PI). Spermatozoa from bulls, boars, rams, rabbits, mice, and men were simultaneously assessed by microscopic examination and flow cytometric evaluation.

The simultaneous examination by fluorescence microscopy coupled to the evaluation of motility revealed that SYBR-14 stained the nuclei of living motile sperm bright green, while non-motile sperm that had lost their membrane integrity stained red due to PI incorporation. A flow cytometry approach was performed by first staining with SYBR-14 and PI and then assessing stain uptake, and determining the proportions of living and dead sperm. Three populations of sperm were identified: SYBR-14 stained living sperm (green), PI stained dead sperm (red), and moribund sperm showed doubly stained. Interestingly, it was observed that the SYBR-14 staining was replaced by PI staining as sperm progressed from living to moribund starting at the posterior region of the sperm head and proceeded anteriorly. A flow cytometry approach coupled to a dual staining of sperm with SYBR-14 and PI readily identified proportions of living and dead sperm in mammalian semen enabling the determination of semen quality.

The DNA content has been determined by flow cytometry in spermatozoa presumed to be the X- and Y- chromosome-bearing gametes from diverse mammals such as bulls, boars, rams and rabbits and even from the spermatozoa of cockerels (4). Dfferences were found in the relative content of DNA and proportions of the presumed X- and Y-sperm populations in cryopreserved semen from bulls by staining sperm nuclei for DNA with the fluorochrome 4'-6-diamidino-2-phenylindole (DAPI). Approximately 5000 stained nuclei were measured for DNA in an epi-illumination flow cytometer. Two symmetrical, overlapping peaks for mammals with X- and Y-spermatozoa were seen whilst only a single distinct peak for cockerel spermatozoa was observed, strongly supporting the interpretation that the peaks represent the X- and Y-sperm populations. The content of DNA in those sperm nuclei nicely correlate with DNA estimates per sperm cell with a difference between the peaks of 3.9, 3.7, 4.1 and 3.9% for bulls, boars, rams and rabbits, respectively. It was found that the X-Y peak (XYp) differences did not vary within animals in each breed, but XYp differences were significant between the breeds. Therefore, these initial studies clearly showed that a flow cytometric approach can be used to assess the proportion of X- and Y-spermatozoa in the semen of domestic animals permitting the development of enrichment techniques for X- or Y-spermatozoa and the determination of the quality and effectiveness of the process (Figure 16).

Bovine spermatozoa were also stained with Hoechst 33342 and analyzed by Flow microfluorometry (5). A bimodal XYp profile was identified with the motile spermatozoa, confirming the observation of Garner *et al*. By diverse sorts coupled to cytometric re-analysis, and orientation experiments, it was suggested that there are indeed two distinct populations of motile spermatozoa.

Dual parameter flow cytometry was also used to investigate cellular changes on the basis of ploidy level, RNA content, and chromatin structure by cell staining with the metachromatic fluorochrome acridine orange, during normal postpartum maturation of germinal male mice tissue. It was found by Janka *et al.* that intensities of red and green fluorescence staining reflect amounts of single- and double-strand nucleic acid sites available for acridine orange (6). Also this approach allowed the identification of testicular, cauda epididymis and vas deferens cell suspensions as well as the determination of the sequential changes in haploid, diploid and tetraploid cell types during the first round of spermatogenesis, and spermatid morphological changes in the same maturation period.

These studies demonstrated that flow cytometric analyses of sperm provided the capability of performing quantitative DNA measurements of the changes in specific fluorescent stained populations of spermatozoa. The differences in the DNA content of X- and Y- chromosome-bearing gametes stained with Hoechst 33342 determined by the bimodal XYp allowed Johnson *et al.* to separate with a flow cytometer/cell sorter approach intact, viable X and Y chromosome-bearing sperm populations of the rabbit that ultimately enabled sex preselection. (7)

Sperm from mice and men were stained with SYBR-14 and PI in order to determine the quality of the sperm and the viability of the spermatozoa. In addition, DNA content was performed following PI staining of spermatozoa (Figure 17). Parallel comparative assays were performed with diploid and tetraploid nuclei controls of chicken erythrocyte nuclei and calf thymocyte nuclei that enable the setting of the cell cycle analysis of the sperm samples (Nunez R. manuscript in preparation 2000)

References
1. Garner DL, Pinkel D, Johnson LA. 1986. Assessment of spermatozoal function using dual fluorescent staining and flow cytometric analyses. Biol. Reprod. 34: 127-138.
2. Graham JK, Kunze E and Hammerstedt RH. 1990. Analysis of sperm cell viability, acrosomal integrity, and mitochondrial function using flow cytometry. Biol. Reprod. 43: 55-64.
3. Garner DL and Johnson LA. 1995. Viability assessment of mammalian sperm using SYBR-14 and propidium iodide. Biol Reprod. 53: 276-284.
4. Garner DL, Gledhill BL, Pinkel D, Lake S, Stephenson D, Van Dilla MA

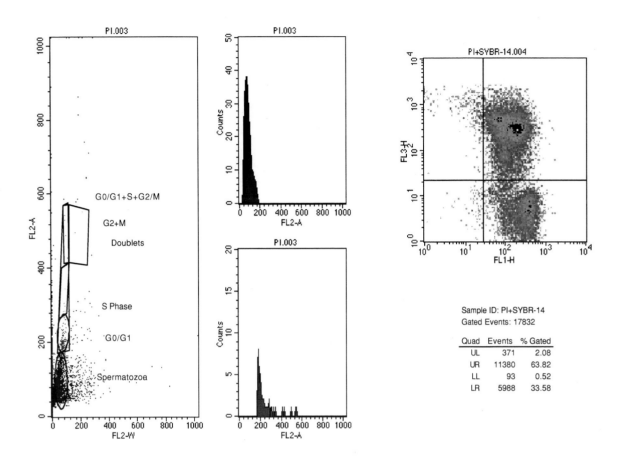

Figure 17. Assessment of the sperm quality by dual staining with SYBR-14 and PI and measurement of DNA content and cell cycle analysis on human sperm. A human sperm sample was evaluated for dual staining (SYBR-14/PI) in order to assess the quality of the sample. Upper right panel shows the dual staining of human spermatozoa. This panel shows a density plot that displays two major clusters of cells. On the lower right quadrant are located the single stained spermatozoa that represent the healthy cells (only SYBR-14 positive). The dual stained spermatozoa (SYBR-14 positive and PI positive) are displayed on the upper right quadrant. They represent the dying sperm cells. On the upper left quadrant are the dead spermatozoa (only PI single positive). The number and % of cells at each quadrant are displayed in the lower right panel. The sample shows that there is a significant number of dying spermatozoa, about 64%. The left panel shows the cell cycle analysis of the sample displayed in the upper right panel. It was found that there is only a minute amount of sperm cells in the diploid region (G0/G1 through G2+M) while the bulk of the cells are located in the subG0 region. The gated cells from the subG0 region are shown in the upper middle panel, while the gated cells containing only singlets from the diploid region are shown on the lower middle panel.

and Johnson LA. 1983. Quantification of the X- and Y-chromosome-bearing spermatozoa of domestic animals by flow cytometry. Biol. Reprod. 28: 312-321.
5. Keeler KD, Mackenzie NM, Dresser DW. 1983. Flow microfluorometric analysis of living spermatozoa stained with Hoechst 33342. J. Reprod.

Fertil. 68: 205-12.
6. Janca FC, Jost LK and Evenson DP. 1986. Mouse testicular and sperm cell development characterized from birth to adulthood by dual parameter flow cytometry. Biol Reprod. 34: 613-623.
7. Johnson LA, Flook JP and Hawk HW. 1989. Sex preselection in rabbits: live births from X and Y sperm separated by DNA and cell sorting Biol. Reprod. 41: 199-203.
8. Nunez R. 2000. Flow cytometric analysis of cell cycle and DNA content on human and mice sperm. Manuscript in preparation.

10

Flow Cytometric Assessment of Allopurinol Susceptibility in *Leishmania infantum* Promastigote

Leishmaniasis is a major tropical and subtropical parasitic disease. The yearly prevalence is estimated at 12 million people world-wide and 200-350 million people are at risk. In the Mediterranean region, leishmaniasis caused by *Leishmania infantum* has emerged as one of the important opportunistic infections of human immunodeficiency virus (HIV) positive individuals (1). Moreover, the prevalence of canine leishmaniasis in this region, may be as high as 42%. Dogs and wild canids are important reservoirs and are mainly responsible for the persistence of the disease in this region (2).

Sodium stibogluconate, N-methyl-D-glucamine antimoniate, amphotericin B, pentamidine and ketoconazole are drugs used in treatment of leishmaniasis. Some of these drugs cause severe adverse side effects and failure of treatment is common. Allopurinol, a purine analogue, has been used for treatment of leishmaniasis, alone or combined with the previously mentioned drugs. In leishmania, allopurinol inhibits purine biosynthesis and hence an inhibition in protein synthesis (3). Low cost, ease of administration (oral), and lack of toxicity make allopurinol a particularly appealing candidate (4).

The effect of allopurinol on wild type promastigotes (*wt-p229*) and on promastigotes of the same isolate which had been cultivated *in vitro* in the presence of up to 800 µg/ml allopurinol for a period of one year (*allo-p229*) was monitored by diverse flow cytometric approaches.

Flow cytometric analysis
The green fluorescence of CFSE, SYBR-14, and FITC and the red fluorescence of PI was excited at 488 nm (FACSCalibur, Becton Dickinson, Heidelberg, Germany). The fluorescence intensities of stained and unstained *wt-p299* and

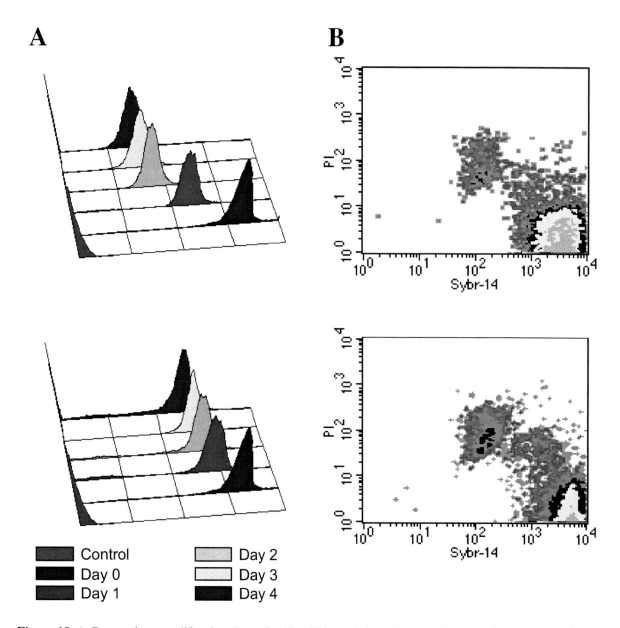

Figure 18. A. Promastigote proliferation determined by CFSE-staining of *wt-p229*. Promastigotes untreated (top panel), or treated with 400 µg/ml allopurinol (bottom panel). B. Promastigote viability. SYBR-14- and PI-staining of *wt-p229*. Promastigotes untreated (top panel), or treated with 400 µg/ml allopurinol (bottom panel).

allo-p229-promastigotes, treated and untreated with allopurinol were determined and compared. At least ten thousand cells were analysed per run and each staining experiment was repeated four times. Data analysis was performed on fluorescence intensities that excluded cell auto-fluorescence and cell debris. CELLQuest analysis software was used to determine the fluorescence and for most of the data analysis, while WinMDI was used to generate the 3 dimensional histogram overlays.

The paper of Kamau *et al.* describes a new technique for the investigation of *Leishmania* promastigote proliferation using CFSE stain, which permits the resolution and tracking of a population of cells that has undergone a different number of cell divisions. Until now incorporation of (^{3}H-thymidine has been the most common method for determining cell division. The use of CFSE assessed the effect of allopurinol on the proliferation of these promastigotes at different time intervals (Figure 18 A). This new flow cytometric application on *Leishmania* opens up potential studies in anti-leishmanial drug pharmacokinetic and toxicology studies (6).

The combination of SYBR-14 and PI have been used extensively in sperm viability studies (5). However, this is the first time that this dual-staining has been used to determine the viability of parasites. The nucleus and mitochondrial DNA of the live promastigotes were stained brilliant green by SYBR-14, while the the nucleus of the dead promastigote was stained red with PI. Some of the promastigotes were dual-stained with a resulting yellowish colour of the nucleus. Flow cytometry was effective in quantifying the resultant fluorescent populations, SYBR-14-stained, PI-stained, and dual-stained promastigotes. Both dyes label DNA thus avoiding the ambiguity of stains that target separate cellular organelles (5). The significant increase of dead (PI-stained) and dying (PI- and SYBR-14-stained) cells in the *wt-p229* promastigotes after exposure to allopurinol indicates a clear allopurinol susceptibility of these promastigotes. In contrast, the proportions of dead and dying cells in the *allo-p229* promastigotes was not significantly influenced by drug exposure (Figure 18 B).

This staining method has the advantage of being rapid (30 min) and, in addition, the cells did not required extra processing prior to the staining. Thus, the study of Kamau *et al.* demonstrated that SYBR-14 when used in combination with PI, was effective for simultaneously visualizing both the living and dead population of *Leishmania* promastigotes before and after treatment with allopurinol (6). The flow cytometry approach permitted the demonstration of differences in allopurinol susceptibility of the two promastigote forms, expanding the spectrum of flow cytometry applications in studies of parasite resistance. The assessment of viability and cellular changes by flow cytometry proved to be a promising way of evaluating the susceptibility and resistance of *Leishmania* promastigotes to allopurinol. The successful application of flow cytometry to determine cellular changes in *Leishmania* cells further opens up future perspectives in determination of effects of anti-leishmanial compounds.

The CFSE assay and the viability assay with SYBR-14 and PI are new tools in the flow cytometric measurement of cell toxicity in parasitology. The flow cytometry approaches enable the detection, differentiation and quantification of cellular changes in these parasites as a result of allopurinol treatment. In addition, they also demonstrate differences in allopurinol susceptibility of the two promastigote forms, hence expanding the spectrum of flow cytometry applications in the field of parasitology and in studies of parasite-drug interactions as well as cellular toxicity.

References
1. Montalban C, Calleja JL, Erice A, Laguna F, Clotet B, Podzamczer D, Cobo J, Mallolas J, Yebra M, Gallego M. 1990. Visceral Leishmaniasis in patients infected with human immunodeficiency virus. J. Infect. 21: 261-270.
2. Baneth G, Dank G, Keren-Kornblatt E, Sekeles E, Adini I, Eisenberger CL, Schnur LF, King R, Jaffe CL. 1998. Emergence of visceral Leishmaniasis in Central Israel. Am. J. Trop. Med. Hyg. 59: 722-725.
3. Frayha GJ, Smyth JD, Gobert JG, Savel J. 1997. The mechanism of action of antiprotozoal and antihelmintic drugs in man. Gen. Pharmacol. 28: 273-299.
4. Quellette M, Papadopoulou B. 1993. Mechanisms of drug resistance in Leishmania. Parasitol. Today. 9: 150-153.
5. Garner DL, Johnson LA, Yue ST, Roth BL, Haugland RP. 1994. Dual DNA staining assessment of bovine sperm viability using SYBR-14 and propidium Iodide. J. Androl. 15: 620-629.
6. Kamau S., Hurtado M., Müller-Doblies U., Grimm F., Nunez R. 2000. Flow cytometric assessment of allopurinol susceptibility in *Leishmania infantum* promastigotes. Cytometry. 40: 353-360.

11

Flow Cytometric Assessment of Transduction Efficiency and Vector Cytotoxicity of HSV-1 Amplicon Vectors

Herpes simplex virus type 1 (HSV-1) has many properties that make it a promising gene transfer vehicle: (i) the genome is a linear, double-stranded DNA of ~152 kb (1), (ii) at least 30 kb of foreign DNA can be inserted, (iii) HSV-1 can infect most cell types, both dividing and non-dividing.

Two distinct types of replication-defective HSV-1-based vectors have been developed: i) the recombinant vectors in which one or more virus genes are replaced with the transgene(s); and ii) amplicons which carry the transgene(s) and two non-coding, *cis*-acting HSV-1 signals, in particular an origin of DNA replication (*ori*) and a DNA cleavage/packaging (*pac*) signal (2). However, these vectors depend on HSV-1 helper functions for replication and packaging into virions.

A first generation helper virus-free packaging system utilized a set of cosmids that represents the HSV-1 genome, with *pac* signals deleted, in five overlapping clones (3). Following transfection into mammalian cells, the cosmids can form a circular HSV-1 genome via homologous recombination and provide all the helper functions required for replication and packaging of co-transfected amplicon DNA. However, in the absence of *pac* signals the reconstituted HSV-1 genomes are packaging-defective and, therefore, do not give rise to progeny virus. The resulting amplicon stocks can efficiently transduce many different cell types (3).

A second generation helper virus-free packaging system has been generated by re-assembling the five cosmid clones, with *pac* signals deleted, in a single

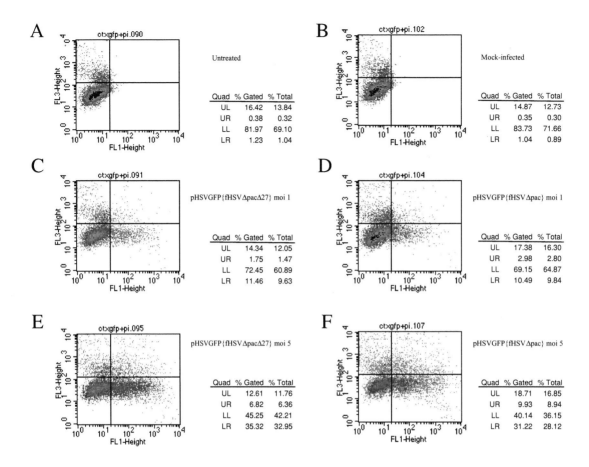

Figure 19. Flow cytometry assessment of transduction efficiency and cytotoxicity of two amplicon vector stocks. Dual fluorescence (GFP- and PI) on untreated cells, mock-infected cells, and cells infected with either pHSVGFP{fHSVΔpacΔ27} or pHSV{fHSVΔpac}) at 1 or 5 moi were evaluated at day five by flow cytometry. Panel A: Untreated cells. Panel B: Mock-infected cells. Cells infected with pHSVGFP{fHSVΔpacΔ27} at a moi of 1 (Panel C) or 5 (Panel E). Cells infected with pHSVGFP{fHSVΔpac} at a moi of 1 (Panel D) or 5 (Panel F). Quadrant values (% of gated and % of total cells) are shown in a Table to the right of each density plot. Cells in the upper left (UL) quadrant (Quad) are GFP-negative and PI-positive (dead cells). Cells in the lower left (LL) quadrant are GFP-negative and PI-negative (non-transduced, live cells). Cells in the upper right (UR) quadrant are GFP-positive and PI-positive (transduced, dead cells). Cells in the lower right (LR) quadrant are GFP-positive and PI-negative (transduced, live cells).

E. coli F-plasmid-based bacterial artificial chromosome (fHSVΔpac) (4). While the formation of a replication- and packaging-competent virus genome from the cosmid clones requires six recombination events, a single recombination event between fHSVΔpac DNA and amplicon DNA is sufficient. To counter this problem, the HSV-1 immediate-early (*IE*) 2 gene which encodes the ICP27 protein has been deleted from fHSVΔpac (fHSVΔpacΔ27; Seaki *et al.*, unpublished data). Although, the formation of a packaging-competent HSV-1 genome is also possible via homologous recombination between fHSVΔpacΔ27 DNA and amplicon DNA, the resulting recombinant virus progeny are replication-defective, as ICP27 is absolutely essential for virus growth. Consequently, cytotoxic effects associated with amplicon vector-mediated gene transfer should be reduced.

Two herpes simplex virus type 1 (HSV-1)-based amplicon vector stocks prepared by transient co-transfection with two different BAC-cloned packaging-defective HSV-1 helper genomes, fHSVΔpacΔ27 and fHSVΔpac, were assayed by flow cytometry, with respect to transduction efficiency and cytotoxicity. Both, vectors are packaging defective because the *pac* signals have been deleted; in addition fHSVΔpacΔ27 contains a deletion in the HSV-1 ICP27 gene which increases the safety of the system (5).

HSV-1 amplicon (pHSVGFP) was packaged into virus particles by transient co-transfection with either fHSVΔpacΔ27 or fHSVΔpac DNA and yielded the vector stocks pHSVGFP{fHSVΔpacΔ27} and pHSVGFP{fHSVΔpac}. Thus, the use of amplicons (fHSVΔpacΔ27 or fHSVΔpac) that express GFP allowed the assessment of the transduction efficiency and potential toxicity of the vectors by flow cytometry in living cultures.

Transduction efficiency
It was determined that both vector stocks, pHSVGFP{fHSVΔpacΔ27} and pHSVGFP{fHSVΔpac} efficiently transduced the target cells. However, the number of green fluorescent cells (GFP-cells) correlated with the multiplicity of infection (moi). In addition, it was found that the highest mean fluorescence intensities were determined at day one after infection, while the highest number of GFP-cells were found at day three after infection.

Vector cytotoxicity
The incubation of the cell populations with propidium iodide (PI) prior to flow cytometric determination allowed the simultaneous assessment of both GFP-positive and cytopathic cells. PI stains genomic DNA in cells that have lost membrane integrity and is, therefore, a reliable marker for cell damage and cell death. It was found that the number of dead cells increased dramatically in all the cultures (transduced or non-transduced) between days three and four after infection. Interestingly, at day five only the cultures infected with a high dose of pHSVGFP{fHSVΔpac} displayed a significant further increase in the number of dead cells (Figure 19).

Flow cytometry allowed the comparison of transduction efficiency and cytotoxicity mediated by the two different amplicon vector stocks. Cultures infected with pHSVGFP{fHSVΔpacΔ27} were found to be clearly more viable than those infected with pHSVGFP{fHSVΔpac} (Figure 19). This study expands the applications of flow cytometry into the gene transfer and vector design fields because it describes the simultaneous assessment of transduction efficiency and cytotoxicity of a vector. Flow cytometry has proven a fast and reliable approach to assess the quality of potential gene transfer vectors prior to their use in preclinical and clinical trials (5).

References
1. Perry LJ, McGeoch DJ. 1998. The DNA sequences of the long repeat region and adjoining parts of the long unique region in the genome of herpes simplex virus type 1. J. Gen. Virol. 69: 2831-46.
2. Spaete RR, Frenkel N. 1992. The herpes simplex virus amplicon: a new eucaryotic defective-virus cloning-amplifying vector. Cell. 30: 295-304.
3. Fraefel C, Song S, Lim F, Lang P, Yu L, Wang Y, Wild P, Geller AI. 1996. Helper virus-free transfer of herpes simplex virus type 1 plasmid vectors into neural cells. J. Virol. 70: 7190-7.
4. Saeki Y, Ichikawa T, Saeki A, Chiocca EA, Tobler K, Ackermann M, Breakefield XO, Fraefel C. 1998. Herpes simplex virus type 1 DNA amplified as bacterial artificial chromosome in *Escherichia coli*: Rescue of replication-competent virus progeny and packaging of amplicon vectors. Hum. Gene Ther. 9: 2787-94.
5. Nunez R., Ackermann M, Saeki Y., Fraefel C. 2000. Flow cytometry assesment of transduction efficiency and cytotoxicity of Herpes Simplex Virus type1 (HSV-1)-based amplicon vectors. Cytometry. In press.

12

Immunophenotyping of DC by Flow Cytometry and Description of Diverse Functional Studies

Flow cytometric assessments on human dendritic cells (DC) of surface markers

The characterization of surface markers on human DC has been a very difficult and elusive task because of the lack of appropriate reagents with high specificity for DC identification (1). However, some molecules whose genes have been cloned and sequenced recently (*e.g.* CD83, DEC-205) have been found to be strongly associated with DC (2,3). Also, a panel of monoclonal antibodies (*e.g.* CMRF-44) that recognize molecules on DC has been raised (4). Therefore, there is a growing need to establish a common and comprehensive nomenclature for such known molecules and for the new monoclonal antibodies, as well as to clarify and define the lineage(s) of DC and the existence of DC subsets. Thus, a flow cytometry approach was instigated in order to evaluate the diverse mAb submitted against two DC populations.

Two DC populations were assessed. Dermal DC like DC (iDC) were generated by culturing PBMC with GM-CSF+IL-4 (5). TGF-β1 was added to half of the culture in order to generate mature Langerhans cell like DC (LC mDC) (6).

In addition, a flow cytometry approach was instigated to assess the DC reactivities. Cells (10^5 cells/100µl) were incubated with 10 µl of mAb on ice for 30'. Cells were washed, two times with wash buffer (Becton Dickinson, San Jose Ca), and stained with 10 ul (1:100 antibody dilution) anti-mouse Ig FITC-labeled on ice for 30'. The cells were washed once and fixed with cellfix (Becton Dickinson, San Jose Ca). Thereafter, cells were acquired (10^4 cells per sample) and analyzed in a FACSCalibur (Becton Dickinson, San Jose Ca) equipped for multiparametric and multicolor analysis with two lasers. One 488 nm Argon laser was used for measurement of forward light scatter (FSC),

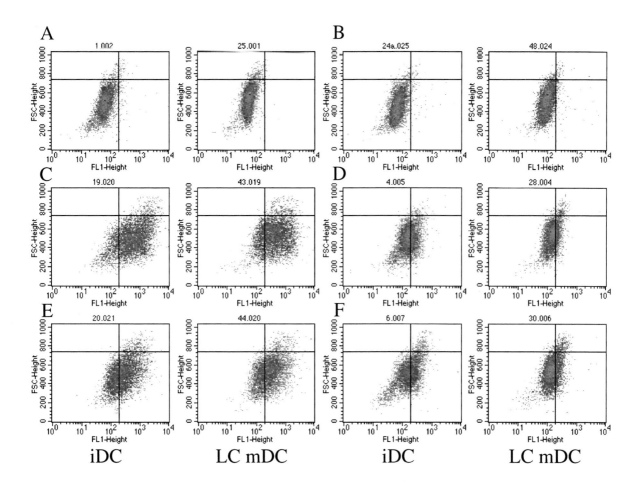

Figure 20. Flow cytometry density plots of representative mAb reactivities against iDC and LC– at 7 days of culture.

orthogonal scatter (SSC) and FITC. Cells were acquired and gated by FSC and SSC. The gated cells were analyzed by (i) histogram plots displaying the fluorescent reactivity collected in Fluorescence 1 (green channel); (ii) Density plots displaying the fluorescent reactivity collected in Fluorescence 1 (green channel) against the FSC. Data and fluorescent signals were collected and stored as list mode files. Flow cytometry measurement on non-stained cells, or on those stained only with the second antibody (2nd Ab), were performed and used as control populations. Cells stained only with the 2nd Ab at day 7 served to determine the markers and the quadrant borders. At least 99% of these cells were located in the lower left quadrant (negative). Data were analyzed with CellQuest (Becton Dickinson, San Jose Ca).

The monoclonal antibody (mAb) reactivities against DC were assessed by

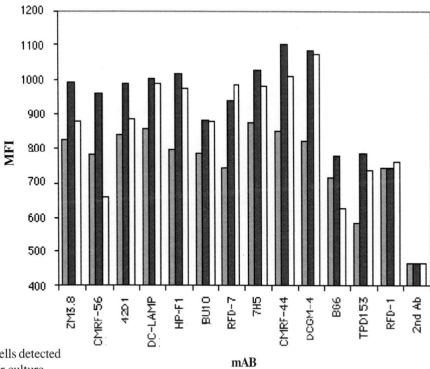

Figure 21. MFI of positive cells detected by mAb against HDCL after culture (no cytokine ▓, TNFα ■, IFNα □)

determining the % of reactivity and mean fluorescent intensity (MFI) of the two DC populations at day 7, 14, 21 and 28.

The simultaneous evaluation on iDC and LC- of FSC and mAb reactivity at four time points allowed the demonstration of the presence of significant variability within the iDC and LC- populations (Figure 20). At day seven, there were several mAb that showed strong reactivity and could be grouped together (Figure 20 panel C and E). In addition there was another group of mAb with an intermediate level of reactivity that could be grouped together (Figure 20 panel D and F). However, some mAb showed no reactivity at day seven (Figure 20 panel A) compared with the negative control (Figure 20 panel B). In addition, variations were found in the reactivity of the mAb at diverse time points. Moreover, at day seven it was found that mAb HP-F1 was negative for iDC and positive for LC-, thus discriminating iDC from LC-. Within the panel analyzed, it was found that at day seven there was a mAb that showed reactivity only against iDC with high FSC (mAb TPD153) or against a fraction of iDC (mAb DC-LAMP and 55K-2).

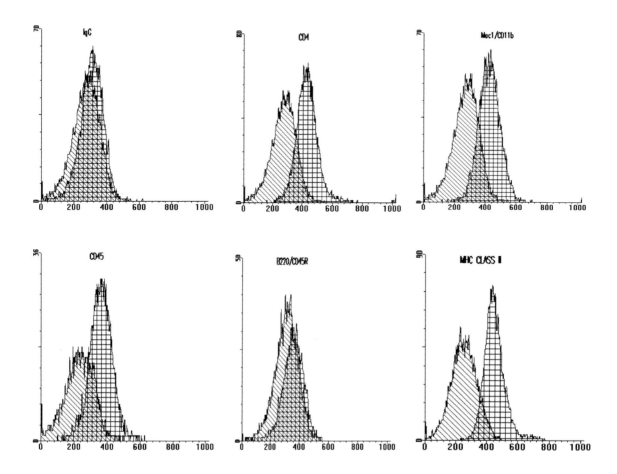

Figure 22. Immunophenotyping of DC by flow cytometry. A mouse DC line was immunophenotyped with a panel of monoclonal antibodies. Histogram of the samples are overlayed with the isotype control.

Thus the flow cytometry approach coupled with the panel of antibodies i) demonstrated the existence of subsets within the iDC and LC- populations; ii) permitted the determination of the kinetics of antigen expression at diverse intervals of time on DC; iii) demonstrated the likelihood of identifying specific markers for subpopulations of DC (7).

Other flow cytometry studies on human DC and human dendritic cell lines (HDCL). Assessing the role of cytokines on the modulation of surface antigen expression on DC

The characterization of surface markers on human DC has been a very difficult and elusive task because of (i) the lack of appropriate reagents such as monoclonal antibodies (mAb) with high specificity for DC identification (1); (ii) the limitations in obtaining pure populations of DC; and (iii) the

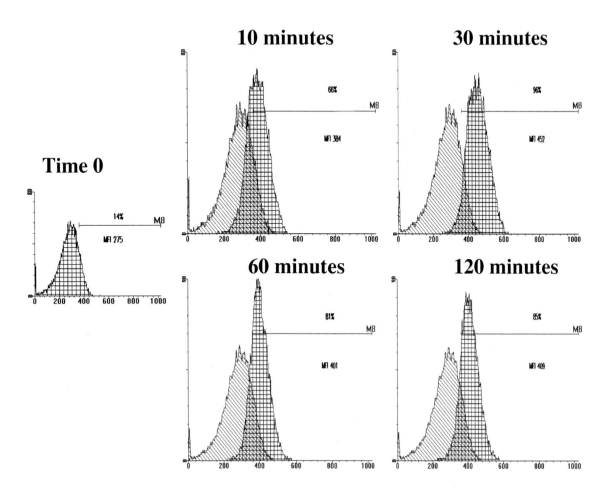

Figure 23. Antigen uptake determined by flow cytometry. Human DC were pulsed with antigen conjugated with fluorescein (OVA-FITC). Thereafter the DC-antigen mixture was incubated at 37°C for various amounts of time. Samples were evaluated at time 0, 10, 30, 60 and 120 minutes. Histograms of the samples are overlayed with the control.

unavailability of homogenous, monoclonal human dendritic cell lines (HDCL) for such studies (1, 5). However, recently technology was developed to immortalize human dendritic cell lines by use of the RAN retrovirus (6) and a panel of mAb with DC reactivity became available. Therefore, a flow cytometry approach was set up and the panel of 13 mAb with DC reactivity was evaluated against a representative HDCL. In addition, the HDCL was cultured with either TNFα or IFNγ in order to determine the influence of these cytokines in the expression of DC antigens detected by the panel of mAb and the likelihood of identifying markers for maturation stages within DC populations.

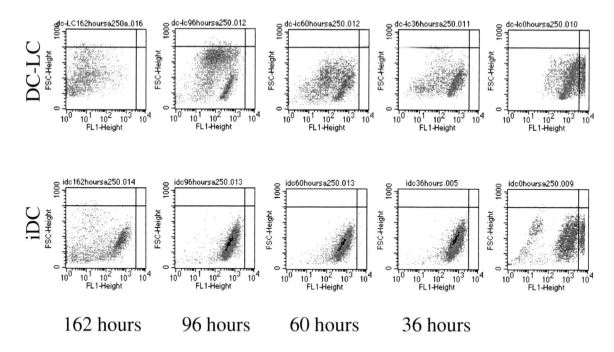

Figure 24. Human DC proliferation determined by CFSE staining. The FACS plots show the decrease in CFSE fluorescence intensity at 0, 36, 60, 96 and 162 hours. Five panels of density plots (upper row) show the CFSE fluorescence for DC under the effect of GM-CSF, IL-4 and TGF-ß1 (DC-LC), while the lower row shows the panels of DC treated with GM-CSF, IL-4 but not TGF-ß1 (iDC).

Dermal DC like (iDC) was generated by culturing adherent PBMC from a healthy donor with (rhGM-CSF+rhIL-4) for 7 days (5). The RAN retrovirus, whose sequence information can be accessed in GenBank under the numbers AF159093, AF159094 and AF159095, was used for immortalization (6). After 7 days of culture, half of the DC cultures were inoculated with RAN retrovirus in order to generate immortalized HDCL. After 21 days, only the cultures with RAN supernatant yielded diverse clones. Only the studies on one clone (HDCL1) are described in this report. The other half of the culture was used as control. HCDL1 was also cultured in the presence of either TNFα or IFNγ.

The results of this study show that 12 out of 13 mAb reacted strongly with untreated HDCL1, confirming that the cell line HDCL1 still shares the expression of markers for DC lineage even after immortalization with RAN virus. Thereafter, HDCL1 was cultured in the presence of either TNFα or IFNγ and flow cytometry evaluation of the TNFα-treated HDCL1 showed (i) an increase in the reactivity of all the mAb compared to the untreated and IFNγ-treated cells, and (ii) that five mAb had MFI >1000 (DC-LAMP, HP-F1, 7H5, CMRF-56 and DCGM-4) (Figure 21). This result suggests that TNFα

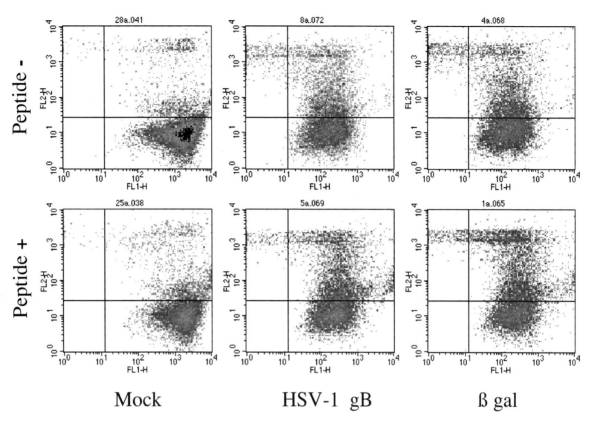

Figure 25. Assessment of *in vitro* CTL activity by flow cytometry. A mouse was immunized with DC transduced with β-gal amplicon and the targets were either MC57 loaded with a β-gal peptide containing a CTL epitope (β-gal) or MC57 loaded with a peptide containing a CTL epitope against gB (HSV-1 gB). The target negative control was MC57 alone (peptide -). A mock immunized mouse was also assayed (Mock).

induced a process of maturation/differentiation coupled to high expression of DC lineage markers.

The treatment of HDCL1 with IFNγ showed that only CMRF-44 and DCGM-4 had MFI >1000, suggesting down regulation in the expression of CMRF-56 and BG6 (Figure 21). Moreover, the EM of the IFNγ-treated HDCL1 demonstrated that the cells died via apoptosis, which was characterized by the typical nuclear changes such as nuclear condensation and fragmentation. In summary one can say, that in contrast to the apoptotic findings detected in the IFNγ-treated HDCL1, TNFα treatment does not induce cell death but an enhanced expression of surface markers. The flow cytometric evaluation on HDCL1 of mAb reactivity demonstrated the presence of significant diversity within the TNFα and IFNγ-treated populations compared with the reactivity of the untreated HDCL1. Thus, the results suggested the existence of diverse stages in the maturation process of DC passing from immature (untreated) to mature (TNFα-treated) and to differentiation towards apoptosis (IFNγ-treated) (7).

Flow cytometric assessments of DC markers and additional functional studies

Dendritic cells (DC) are a complex group of mainly bone marrow derived cells that have been found to play an important role in the afferent branch of the immune response (1). However, DC represent only a minute subpopulation of the peripheral blood mononuclear cells, as well as of bulk cellular populations of the lung, intestine, genitourinary tissue, and lymphoid tissue. In addition, DC have been found in the epidermis, dermis and mucous membranes and constitute about 2% of the total cellular population of the human epidermis (2). The so called Langerhans cells (LC), are skin-derived DC that migrate to the regional lymphoid organs after take up of antigen and undergo an activation/maturation step. Thereafter LC interact and activate T cells. Because of their significant capability to take up soluble antigens, processe and present them to responder cells in the lymphoid tissues in the context of the restricted MHC pathway, LC have been considered one of the most important elements in the afferent arm of the immune response (1-4).

Immunophenotyping of DC by fow cytometry assays

The surface antigens on DC have been studied by diverse flow cytometric approaches. The panel of monoclonal antibodies used for the DC immunophenotyping was PE-, PerCP-, APC- or FITC-conjugated. The negative controls for the FACS were isotype matched FITC-, PE-, PerCP-, or APC labeled non-related antibodies, allowing the setting of the markers in the plots. One or multicolor immunofluorescence analysis was performed after appropriate color compensation for each one of the markers in a flow cytometer. A single color assay allows the overlaying of the histograms obtained from the sample with the one of the isotype control as is shown in Figure 22. The overlay representation has the advantage that it provides a visual assessment of the degree of similarity or difference that the marker on the DC has in comparison with the isotype control.

Antigen uptake determined by flow cytometry

Determination of antigen uptake, one of the most important functions of DC, can also be assayed by flow cytometry. DC is pulsed with antigen conjugated with fluorescein (OVA-FITC, Molecular Probes, Eugene, OR). Graded doses (0.001 to 1mg/ml) of antigen are added to cells kept on ice for 10 min. Thereafter the DC-antigen mixture is incubated at 37°C for various amounts

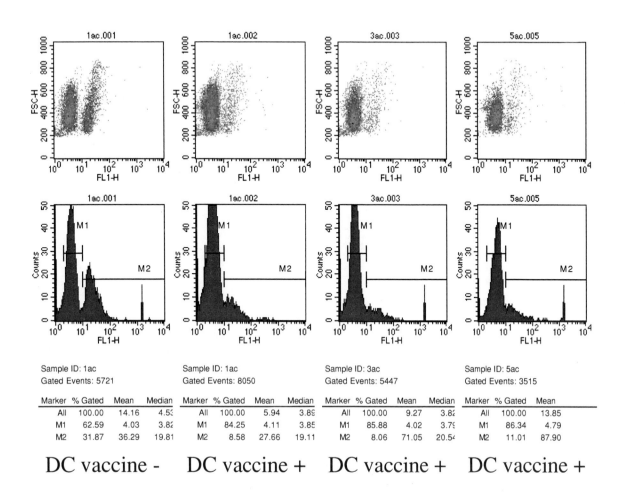

Figure 26. Assessment of *in vivo* CTL activity by flow cytometry. Mice were immunized with DC transduced with rabies amplicon (DC vaccine +) or mock immunized (DC vaccine -). Three months later, all mice were boosted with a commercial rabies vaccine. Four days after boosting, mice were infused with a mixture of target cells consisting of non-infected targets labeled with DIOC3 at low concentration and target cells infected with rabies virus but labeled at high concentration. Twenty four hours later, mice were bled and cells were evaluated by flow cytometry. Upper panels show the density plots of cells located in the gate for target cells. Middle panels show the histograms of the gated cells with markers for the peaks (M1 is the marker for the peak with low concentration of the label and M2 is the marker for the peak with the high concentration of the label). Lower panels show the statistical data for the cells included in the markers.

of time. Samples are evaluated at time 0, 10, 30, 60 and 120 minutes. Control cells incubated with the same amount of antigen are kept on ice for the same period of time. Antigen uptake is stopped by adding ice-cold FACS buffer. The FACS buffer contains PBS supplemented with 5% FCS and 0.01 % sodium azide that inhibits cell metabolism. Subsequently, the cells are extensively washed and fixed with 2% formaldehyde followed by FACS analysis. Data on antigen binding of control cells kept on ice throughout the entire experiment are used as control. Also, in this assay, the overlay representation complemented with the assessment of % of cells positive and the measurement of mean fluorescence intensity (MFI) of the sample provide a clear determination of the DC uptake capability as is shown in Figure 23 (6).

Flow cytometry assessment of DC division under cytokine modulation
The stable intracytoplasmic dye, 5-, 6-carboxyfluorescein diacetate succinimidyl ester (CFSE) was used as a quantitative method to measure cell division allowing i) identification of cell progeny and ii) analysis of the division history of individual cells that have undergone multiple rounds of division seen as discrete peaks with a progressive reduction of CFSE fluorescence intensity (5).

DC proliferation determined by CFSE staining
DC proliferation in the presence and absence of cytokines was determined by CFSE staining and the results from each treatment and time interval were grouped. The FACS plots show the decrease in CFSE fluorescence intensity in subsequent days (Figure 24). Five panels of density plots (upper row) show the CFSE fluorescence for DC under the effect of GM-CSF, IL-4 and TGF-β1 while the lower row shows the panels of DC treated with GM-CSF, IL-4 but not TGF-β1 (Figure 24). The different panels in the DC treated with TGF-β1 (upper row) show a series of discrete peaks exhibiting a progressive serial decrease of CFSE fluorescence, at different time intervals, a feature suggestive of cell division. In contrast the lower row shows that the DC has not decreased in fluorescence after 36 hours. In addition, the treatment with TGF-β1 yields a population with high FSC. Overall, these results suggested that the DC treated with TGF-β1 can divide and differentiate into at least two subsets of DC, while the untreated DC remain in a steady state of differentiation (7).

Assessment of *in vitro* cytotoxity assay by flow cytometry
Flow cytometry assessment of *in vitro* and *in vivo* CTL activity by T cells triggered by DC priming has been described. The assay for *in vitro* CTL uses a two-color cytometry approach with DIOC3 as membrane stains for the target cells, and PI as nuclear stain of damaged or dead target cells. DIOC3 stains green and is detected in FL1. PI stains red and is detected in FL2 and FL3. The assay shown in Figure 25 does not display the unstained effector cells. The effectors can be gated out because they have lower FSC and SSC. The DIOC3-stained live, target cells are located in the lower right quadrant. The dead target cells are located in upper left quadrant. The dying cells are located in the upper right quadrant. The labeling and the CTL procedure are performed

with LIVE/DEAD™ cell cytotoxicity assay kit (Molecular Probes, Eugene Oregon). Target and effector cells were cultured for 4 hours, then PI was added to the cultures and incubated for 30 minutes. The cultures were washed and analyzed in a FACSCalibur.

Thus, this flow cytometric approach was used for assaying the CTL activity generated in mice immunized with DC transduced with the amplicon β-gal (Figure 25). Mouse 5 was immunized with DC transduced with amplicon β-gal and the targets were either MC57 loaded with a β-gal peptide containing a CTL epitope (β-gal) or MC57 loaded with a peptide containing a CTL epitope against gB (HSV-1 gB). The target negative control was MC57 alone (peptide -). A mock immunized mouse was assayed (Mock). In summary, it was found that, there was a negligible (< 15% of antigen specific lysis) CTL activity after immunization of mice with DC transduced with amplicon(s) compared with the Mock mice and between the loaded (peptide +) and unloaded (peptide -) targets (8).

Assessment of *in vivo* cytotoxity assay by flow cytometry
A flow cytometric approach was also used for assaying the *in vivo* CTL activity generated in mice immunized with DC transduced with amplicon carrying the gene for the rabies glycoprotein and boosted with a commercial vaccine four days prior to infusion with a mixture of 10^5 target cells. Two groups of target cells were used and labeled with DIOC3. Moreover, non-infected target cells were labeled at low concentration while target cells infected with rabies virus were labeled at high concentration. Twenty-four hours later, mice were bled and cells were evaluated by flow cytometry (Figure 26). It was expected that the mice primed by the DC transduced with the rabies amplicon could generate a CTL activity that destroyed the targets infected with rabies virus while the non-infected targets could remain. It was found that the three mice immunized with DC cleared out most of the infected targets while the control did not (Figure 26) (8).

References
1. Banchereau J., Steinman RM. 1998. Dendritic cells and the control of immnunity. Nature. 392: 245-252.
2. Bender A., Sapp M, Schuler G., Steinman RM., Bhardwaj N. 1996. Improved methods for the gneneration of dendritic cells from nonproliferating progenitors in human blood. J. Immunol. Methods. 196: 121-135.
3. Jiang W, Swiggard WJ, Heufler C., Peng M., Mirza A, Steinman RM, Nussenzweig MC. 1995. The receptor DEC-205 expressed by dendritic

cells and thymic epithelial cells is involved in antigen presentation. Nature. 375: 151-155.
4. Hock BD, Sterling GC, Daniel PB., Hart DN. 1994. Characterization of CMRF-44, a novel monoclonal anibody to an activation antigen expressed by allostimulatory cells within peropheral blood, including dendritic cells. Immunology. 83: 573-81.
5. Sallusto F., Lanzaveccia A. 1994. Efficient presentation of soluble antigen by cultured human dendritic cells is maintained by granulocate/macrophage colony-stimulating factor plus interluekin 4 and downregulated by tumor necrosis factor alpha. J. Exp. Med. 179: 1109.
6. Geissmann F., Prost C., Monnet J-P., Dy M., Brousse N., Hermine O. 1998. Transforming growth factor beta1, in the presence of granulocyte/macrophage colony-stimulating factor and interleukin 4, induces differentiation of human peripheral blood monocytes into dendritic Langerhans cells. J. Exp. Med. 187: 961-966.
7. Nunez R, Filgueira L. 2000. Flow cytometry assessment of monoclonal antibody (mAb) reactivities against Dendritic cells (DC) In, D. Mason, Editor, Leukocyte Typing VII, Oxford University Press. In Press.
8. Nunez R, Sanchez M., Wild P., Filgueira L., Nunez C. 1998. Characterization of two human dendritic cell-lines that express CD1a, take-up, process and present soluble antigens and induce MLR. Immunol. Lett. 61: 33-43.
9. Nunez R., Filgueira L., Nunez C. 2000. Flow cytometric assessment on human dendritic cell lines (HDCL) of monoclonal antibody (mAb) reactivities and determination of surface antigen expression by cytokine modulation. D. Mason, Editor, Leukocyte Typing VII, Oxford University Press. In Press.
10. Steinman, R.M. 1991. The dendritic cell system and its role in immunogenicity. Ann. Rev. Immunol. 9: 271-96.
11. Katz, S. I., Cooper, K. D., Iijima, M. and Tsuchida, T. 1985. The role of Langerhans cells in antigen presentation. J. Invest. Dermatol. 85: 96-98.
12. Cella, M., F. Sallusto, and A. Lanzaveccia. 1997. Origin, maturation and antigen presenting function of dendritic cells. Curr. Opin. Immunol. 9: 10-6.
13. Kamau S., Hurtado M., Müller-Doblies U., Grimm F., Nunez R. 2000. Flow Cytometric assessment of allopurinol susceptibility in *Leishmania infantum* promastigotes. Cytometry. 40: 353-360.
14. Nuñez R., Fraefel C., Zanoni R., Brudner L., Suter M., Ackermann M. 2000. Development of a cellular vaccine against viral or soluble antigens consisting of dendritic cells transduced with HSV-1 amplicon vectors. Manuscript in preparation.

13

Summary

Flow cytometry is a methodology for determining and quantitating cellular features, organelles or cell structural components primarily by both optical and electronic means. Although it measures one cell at each time, the newest instrumentation is capable of processing hundreds of thousands of cells in a few seconds. Flow cytometry can be used to count and even distinguish cells of different types in a mixture by quantitating their structural features. Therefore, flow cytometry has great advantages compared to traditional microscopy since it permits the analysis of a greater number of cells in a fraction of the time. In addition cell sorting with flow cytometers has been a powerful tool for diverse fields in research and clinical applications.

Since the early 1970's, flow cytometers that do not employ fluorescence have been commercially available. They were initially used for complete blood cell counts in clinical laboratories. Their ease of handling and reliable results increased and popularized their use since then. The newest and most versatile research instruments employ fluorescence and are named flow cytofluorometers. The world-wide utilization of flow cytometry is easily demonstrated by the use of flow cytometric data in almost any issue of a cell biology journal and in a large percentage of papers published in journals in the fields of biomedical science and immunology. About 43200 citations containing flow cytometry data has been compiled to date by MEDLINE (July 11th, 2000).

Flow cytometers are widely found in all leading biomedical research institutions and universities, where they are used for performing tasks that require analytical precision and high throughput. Flow cytometers also have a key role in hospital and medical centers worldwide, where they are widely used for diagnosis as well as research. There are several thousand flow cytofluorometers in clinical use worldwide. The major diagnostic applications being ploidy, cell cycle and surface analysis of cancers. They are also of use in the study of surface markers of lymphomas and leukemias which are of diagnostic and prognostic value. Flow cytometry has been the method of choice for monitoring the progression of AIDS and the response to treatment by measuring CD4 lymphocyte levels in the blood. Less expensive alternative technologies are not yet available for performing such tasks in clinical and research settings. In addition, sorting and high speed sorting are becoming increasingly important in the performance of research, clinical trials, clinical applications and teaching.

In this report, several special applications of flow cytometry are described and examples are included. Moreover, the role of flow cytometry in the immunophenotyping and functional characterization of dendritic cells is stressed.

A study on sperm from mice and men is described in which spermatozoa were stained with SYBR-14 and PI in order to determine the quality of the sperm and the viability of the spermatozoa. In addition, DNA content analysis was performed following PI staining of spermatozoa. Parallel comparative assays were performed, with diploid and tetraploid nuclei controls of chicken erythrocyte nuclei and calf thymocyte nuclei, that enabled the calibration of cell cycle analysis of the sperm samples.

Another study dealt with the capability of a new flow cytometry approach to demonstrate differences in allopurinol susceptibility of two promastigote forms, expanding the spectrum of flow cytometry applications into studies of parasite resistance. The assessment of viability and cellular changes by flow cytometry proved to be a promising way of evaluating the susceptibility and resistance of leishmania promastigotes to allopurinol. The successful application of flow cytometry to determine cellular changes in *Leishmania* cells further opens up future perspectives in determination of effects of anti-leishmanial compounds. The new assays described in this report (CFSE assay and the viability assay with SYBR-14 and PI) are new tools in the flow cytometric measurement of cell toxicity in parasitology.

The flow cytometry approaches enable the detection, differentiation, and quantification of cellular changes in these parasites as a result of allopurinol treatment. In addition, they also demonstrate differences in allopurinol susceptibility of two promastigote forms, hence expanding the spectrum of flow cytometry applications in the field of parasitology and in studies of parasite-drug interactions as well as cellular toxicity.

In addition, a flow cytometric approach was applied in the assessment of vectors for gene therapy. Thus, flow cytometry allowed the comparison of the transduction efficiency and cytotoxicity mediated by two different amplicon vector stocks. Cultures infected with pHSVGFP{fHSVΔpacΔ27} were found to be clearly more viable than those infected with pHSVGFP{fHSVΔpac}. This study expands the applications of flow cytometry into the gene transfer and vector design fields because it allows the simultaneous assessment of transduction efficiency and cytotoxicity of a vector. Flow cytometry has proven a fast and reliable approach to the assessment of the quality of potential gene transfer vectors prior to their use in preclinical and clinical trials.

In the field of immunology flow cytometry has been used for immunophenotyping and in the assessment of functional characterization of dendritic cells (DC). The surface antigens on DC have been studied by diverse flow cytometric approaches including one or multicolor immunofluorescence analysis. The panel of monoclonal antibodies used for the DC immunophenotyping was PE-, PerCP-, APC- or FITC-conjugated. The negative controls for the FACS were isotype matched FITC-, PE-, PerCP-, or APC-labeled non-related antibodies, permitting the setting of the markers in the plots. The assays were performed after appropriate color compensation for each one of the markers. A single color assay allows the overlaying of the histograms obtained from the sample with the one for the isotype control. Moreover, the overlay representation has the advantage that it provides a visual assessment of the degree of similarity or difference that the marker on the DC has in comparison with the isotype control. Further immunophenotyping assays permitted the determination, in human dendritic cells, of: i) the existence of subsets within the iDC and LC- populations; ii) the kinetics of antigen expression at diverse intervals of time on DC; and iii) the likelihood of identification of specific markers for subpopulations of DC.

Further immunophenotyping assays on human DC and immortalized DC demonstrated the presence of significant diversity within the TNFα and IFNγ-treated populations compared with the reactivity of the untreated HDCL1. Thus, the results suggested the existence of diverse stages in the maturation process of DC passing from immature (untreated), to mature (TNFα-treated) and to differentiation towards apoptosis (IFNγ-treated).

Determination of antigen uptake, one of the most important functions of DC, was also assayed by flow cytometry. DC was pulsed with antigen conjugated to fluorescein (OVA-FITC). Thereafter the DC-antigen mixture was incubated at 37°C for various amounts of time. Samples were evaluated at time 0, 10, 30, 60 and 120 minutes. In this assay, the overlay representation complemented the assessment of % of cells positive and the measurement of mean fluorescence intensity (MFI) of the sample, providing a clear determination of the DC uptake capability.

DC division under cytokine modulation was also determined by flow cytometry. The stable intracytoplasmic dye, 5-, 6-carboxyfluorescein diacetate succinimidyl ester (CFSE) was used as a quantitative method to measure cell division allowing i) identification of cell progeny and ii) analysis of the division history of individual cells that have undergone multiple rounds of division. These are seen as discrete peaks with a progressive reduction of CFSE fluorescence intensity. This study shows the effect of GM-CSF, IL-4 and TGF-β1 on DC compared to those DC treated with GM-CSF, IL-4 but not TGF-β1. The different panels in the DC treated with TGF-β1 show a series of discrete peaks exhibiting a progressive serial decrease of CFSE fluorescence, at different time intervals, a feature suggestive of cell division. In contrast the DC without TGF-β1 does not decrease in fluorescence after 36 hours. In addition, the treatment with TGF-β1 yields a population with high FSC. Overall, these results suggest that the DC treated with TGF-β1 can divide and differentiate into at least two subsets of DC, while the untreated DC remain in a steady state of differentiation.

Flow cytometry assessment of *in vitro* and *in vivo* CTL activity by T cells triggered by DC priming is also described in this study. The assay for *in vitro* CTL uses a two-color cytometry approach with DIOC3 as membrane stains for the target cells, and PI as nuclear stain of damaged or dead target cells. DIOC3 stains green and is detected in FL1. PI stains red and is detected in FL2 and FL3. The DIOC3-stained live target cells are located in the lower right quadrant. The dead target cells are located in the upper left quadrant. The dying cells are located in the upper right quadrant. This flow cytometric approach was used for assaying the CTL activity generated in mice immunized with DC transduced with a β-gal amplicon. It was found that there was a negligible (< 15% of antigen specific lysis) CTL activity after immunization of mice with DC transduced with amplicon(s) compared with the Mock mice and between the loaded (peptide +) and unloaded (peptide -) targets.

Furthermore, an additional flow cytometric approach was used for assaying the *in vivo* CTL activity generated in mice immunized with DC transduced with amplicon carrying the gene for the rabies glycoprotein and boosted with a commercial vaccine four days prior to infusion with a mixture of 10^5 target cells. Two groups of target cells were used and labeled with DIOC3. Non-infected target cells were labeled at low concentration while target cells infected with rabies virus were labeled at high concentration. Twenty-four hours later, mice were bled and cells were evaluated by flow cytometry. It was expected that the mice primed by the DC transduced with the rabies amplicon could generate a CTL activity that destroyed the target cells infected with rabies virus, while the non-infected target cells could remain. It was found that the three mice immunized with DC cleared out most of the infected targets while the control did not.

This book describes the field of cytometry and its importance in biomedicine and bioscience. In particular, it describes some new and inportant applications in the areas of immunology, parasitology, gene therapy, andrology and hematology.

14

Protocols

This section includes some protocols for cytometry technics kindly provided by the Salk Flow Cytometry home page http://pingu.salk.edu/fcm/protocols.html and from David Chambers (for coupling and staining procedures) reproduced here with permission of D. Chambers.

Fluorescein labeling of proteins
This method uses carboxyfluorescein succinimidyl ester (CFSE) rather than fluorescein isothiocyanate, resulting in more reliable labeling. Succinimidyl esters are excellent reagents for amine modification since the amide products formed are very stable. CFSE has high reactivity with aliphatic amines, low reactivity with aromatic amines, including tyrosine.

Reagents
- CFSE Carboxyfluorescein succinimidyl ester (Molecular Probes C-1311). Store desiccated at -70 °C
- DMSO Dimethylsulfoxide, anhydrous
- PBS Phosphate buffered saline, pH 7.4
- Protein to be labeled, purified, ~1 mg/ml in PBS
- Column for gel filtration, *e.g.* 10 ml Sephadex G-10 column
- Dialysis tubing, 10,000 MW cutoff
- Centricon microconcentrators (Amicon)

Method
1. Prepare or otherwise obtain pure protein; make sure it is free of other contaminating proteins (*e.g.* albumin).
2. Ensure protein to be labeled is in a suitable buffer. Buffers containing TRIS are NOT acceptable since the TRIS interferes with labeling. A reasonable buffer to use is PBS, pH 7.4. If necessary, exchange current buffer for PBS using one of three methods:
a. Gel filtration. Do not use if you have less than 1 mg protein.
b. Dialysis. Microdialysis is probably the best method if you do not have very

much protein.

c. Centricon microconcentrators.

3. Adjust protein concentration to ~ 1 mg/ml.
4. Prepare CFSE, 1.5 mg/ml in anhydrous DMSO. CFSE is EXTREMELY moisture sensitive! Store in desiccator at –70 °C, allow desiccator to warm to room temperature before opening. Dilute CFSE in anyhdrous DMSO immediately before use.
5. Add 100 µl CFSE per 1 ml of protein solution.
6. Incubate for 90 minutes in the dark at room temperature with continuous gentle agitation. Alternatively, incubate overnight in the dark at 4 °C with continuous gentle agitation.
7. Exact conjugation efficiency depends on temperature, length of incubation, concentration of protein, concentration of CFSE, and nature of protein. For critical applications, conjugate and test a small amount first to verify the conditions.
8. Separate labeled protein from free fluorescein compounds by extensive dialysis versus PBS or by gel filtration. Concentrate with Centriprep and/or Centricon concentrators as necessary.
9. Microfuge on high for 10 minutes and filter through 0.22 µm filter. Optionally add 0.5% azide as preservative.
10. Determine concentration and F/P ratio by measuring absorbance at 280 nm and at 495 nm. For antibodies, use the following formulae to get approximate values:
Protein concentration (mg/ml) = (OD280 - 0.35 x OD495) / 1.4/P ratio = (3.3 x OD495) / (OD280 - 0.35 x OD495).

Phycoerythrin conjugation protocol

David Chambers's method modified from references (2) and (3). This method has been used to conjugate a mouse IgG2a monoclonal antibody (IB4). It works well.

Reagents
- Ab Antibody, about 2.5 mg/ml.
- PE R-Phycoerythrin, purified (Molecular Probes P-801 in 60% ammonium sulfate).
- PBS Phosphate buffered saline without Ca^{2+}/Mg^{2+}.
- MeOH Methanol.
- DMSO Dimethyl sulphoxide, anhydrous.
- SPDP 3-(2-pyridyldithio)propionic acid N-hydroxysuccinimide ester (Sigma).
- SMCC Succinimidyl trans-4-(N-maleimidylmethyl)cyclohexane-1-

carboxylate (Molecular Probes).
- DTT Dithiothreitol.
- NEM N-ethyl maleimide.
- 10 ml Sephadex G-10 columns (2).
- Separation column. A long Sephacryl S-300 column. *e.g.* 117 x 1 cm. You can't really get away with anything smaller. Use a long thin one rather than a short fat one. Separation of the conjugate from the free reagents is tricky.
- Centricon 30 and Centriprep 30 centrifugal concentrators (Amicon).
Note: Room temperature is considered to be 22 °C.

Method
1. Preparation of PE. R-PE from Molecular Probes. Part number P-801, R-PE in 60% NH4SO4 at 4 mg/ml. 0.25 ml of P-801. Spin 10 mins on low in microfuge. Discard supernatant. Resuspend PE in 0.25 ml PBS. Dialyze twice for 30 mins versus 150 ml PBS at room temperature.
2. SPDP modification of PE. Adjust PE to approx 1.4 mg/ml (=0.7 ml) in PBS. Add 16 µl SPDP (1.3 mg/ml in MeOH). Incubate at room temperature for 2.5 hours

Note: You must time steps 3 and 4 to finish at the same time, since the reaction products are unstable.

3. SMCC modification of antibody. 1 ml purified antibody at approx 2.5 mg/ml in PBS. Antibody must be fairly pure *i.e.* purify on protein A or G column before conjugating. Add 20 µl SMCC, 1.7 mg/ml in DMSO. Incubate for 1 hour.
4. DTT treatment of SPDP-PE. Take product from step (2) and add 30 µl DTT (0.5M, *i.e.* 77 mg/ml in PBS). Incubate for 30 mins at room temperature.
5. Purification of reactants. Separate derivatized PE and Ab from activators on G-10 columns. You should now have approximately 2 ml of each reactant.
6. Conjugation. Mix derivatized reagents and incubate at 4 °C overnight on rotary mixer.
7. Stop reaction. Add 80 µl N-ethyl maleimide (0.1 mM) to reaction mixture to stop reaction. Agitate and incubate 30 mins at room temperature
8. Concentrate product. Concentrate to 0.5 ml or less on Centricon concentrator (Centricon 30 or 100 are fine).
9. Separate conjugate. Apply to top of separation column. Elute with PBS. Pump column at approximately 0.3 ml/min. Order of elution is conjugate first (MW ~400,000) then free PE (MW ~240,000) then free Ab (MW ~150,000). Begin collecting 1.0 ml fractions when pink band approaches bottom of column and continue for 10 fractions after it has disappeared.

10. Evaluate separation. Measure OD280 and OD565 - the former is protein, the latter the PE absorption peak. Test activity of each fraction against known reactive cells by flow cytometry. Pool these fractions.
11. Final concentration. Concentrate pooled conjugate fractions using Centriprep and Centricon concentrators. Add 0.5% sodium azide. Store at 4 °C protected from light. Do not freeze.

Paraformaldehyde preparation

Mary-Ann Campbell's method. This is a quick method of getting paraformaldehyde (PFM) into solution. It is a good alternative to the traditional methods which can take hours, and has the advantage that it is so quick that you are more inclined to prepare fresh PFM each time.
CAUTION! Flammable. Irritant vapor. Probably carcinogenic.

Method
1. 0.1 g paraformaldehyde powder in a small glass tube.
2. Add 0.5 ml distilled water.
3. Add 1 drop 0.5 - 1.0 M sodium hydroxide.
4. Heat to approx 80 °C for 2-3 mins; shake in water bath until PFM has dissolved. A beaker of very hot water will do.
5. Add 9.5 ml PBS. Correct pH with HCl if necessary.
6. Makes 10 ml of 1% PFM solution.

Flow cytometric determination of leukocyte surface antigens in whole blood

This method assumes that you have a mouse IgG monoclonal antibody directed to the antigen you wish to quantitate and that you have fluorescein labeled and fully characterized this antibody.

Quantitation of cell surface antigens in whole blood with the flow cytometer is very simple:
1. Collect blood
2. Add antibody
3. Calibrate the flow cytometer
4. Make the measurements

Reagents
- PBS (Phosphate buffered saline) pH 7.4.
- Saponin (Sigma) - make 10 mg/ml saponin in PBS. Add azide and store in refrigerator.
- Blood.
- Antibody.

Collect blood
1. Collect blood by standard venipuncture into an appropriate anticoagulant. Use EDTA or heparin (10 iu/ml). Do not use citrate, this produces an acidic environment which will quench your fluorescein labeled antibody!
2. Collect blood with a large gauge needle into a syringe. Do not subject the sample to large stresses by pulling too hard on the syringe. If you do this you will certainly damage erythrocytes and probably also damage leukocytes.
3. You may collect the sample directly into anticoagulant in the syringe. If you collect in a syringe without anticoagulant, carefully introduce the blood into a tube containing anticoagulant. Do not allow the sample to froth in either case.

Note: HOWEVER YOU COLLECT THE BLOOD, ENSURE THAT IT IS COMPLETELY MIXED WITH ANTICOAGULANT. MIX BY INVERSION AT LEAST 10 TIMES!

4. If you are conducting a study of patient groups it is desirable to draw samples from matched control subjects at the same time.
5. Try to make sure that all blood is drawn at a similar time of day; if not possible, record the time of day it was drawn and use an appropriate control subject.
6. If you are interested in "resting" levels of antigens, immediately put the tube of blood into ice. KEEP COLD THROUGHOUT.

Add antibody to the blood
1. Divide the blood into the appropriate aliquots and add labeled mAb to each aliquot as desired. Make sure that the final mAb concentration is at saturation - you may need to determine this by titration. Do blocked samples as necessary.
2. Incubate 30 minutes on ice.

Note that this is for directly labeled antibody. If you need to use an indirect technique, you will need to do washes and incubate with secondary reagent at this point.

Make measurements
1. Prepare a stock solution of saponin, 10 mg/ml in PBS. Add 0.1% sodium azide and store in the refrigerator.
2. Before the measurement phase of the experiment starts, prepare a sufficient quantity of tubes each containing 0.05 ml saponin solution. Place these on ice and allow to cool.
3. Add 0.025 ml blood to a 0.05 ml aliquot of saponin and mix with the pipette tip. Wait 5 seconds then add 0.45 ml PBS.

4. Periodically agitate for a short period, up to 30 seconds, monitoring sample clarity visually. It is easy to tell when the erythrocytes lyse, because the sample clears rapidly. As soon as lysis has occurred, return the tube to ice. Allow a further 10 seconds to complete the lysis and read on the flow cytometer. Use characteristic forward versus orthogonal light scattering to set a gate for the type of cells in which you are interested. Acquire at least 5000 events and store pending off-line analysis.
5. You can add up to 0.05 ml blood to each saponin aliquot with good results. Do not add more or the erythrocytes will not lyse properly.
6. It seems that human leukocytes are quite resistant to saponin lysis, and the samples will be stable on ice for about 3-5 minutes.
7. Other species differ in this respect and you may need to modify the lysis protocol to suit. Good results are obtained with human, rabbit, rat, baboon and pig (but not hamster). For optimum leukocyte stability it may be necessary to keep the sample cold while it is being analysed on the cytometer (use an ice bath).
8. The concentration of saponin used here does not appear to remove antigens from the cell membrane, however test this with purified cells if you are suspicious.

References
1. Chambers JD, Simon SI, Berger EM, Sklar LA, Arfors K-E. 1993. Endocytosis of beta2 integrins by stimulated human neutrophils analyzed by flow cytometry. J. Leukocyte Biol. 53: 462-469.
2. Kronick MN, Grossman PD. 1983. Immunoassay techniques with fluorescent phycobiliprotein conjugates. Clin. Chem. 29: 1582-1586.
3. Oi VT, Glazer AN, Stryer L. 1982. Fluorescent phycobiliprotein conjugates for analyses of cells and molecules. J. Cell Biol. 93: 981-986.

Appendix 1

Commercial Resources for Flow Cytometry on the Web

See also: Salk Flow Cytometry Lab Bioscience Links at http://pingu.salk.edu/fcm/sites.html

http://www.abp.com/
http://www.augusta.co.uk/aberinstruments/
http://www.enterprise.net/appcysys/index.htm
http://www.amersham.co.uk/life/
http://www.bangslabs.com/index.html
http://pingu.salk.edu/fcm/sites.html
http://www.biodesign.com/
http://pingu.salk.edu/fcm/protocols.html
http://www.biomeda.com/

Appendix 2

Abbreviations used in Flow Cytometry

3-(2-pyridyldithio)propionic acid N-hydroxysuccimidine ester (SPDP)
4'-6-diamidino-2-phenylindole (DAPI)
5-, 6-carboxyfluorescein diacetate succinimidyl ester (CFSE)
Allophycocyanin (APC)
Becton Dickinson (BDIS) (BD)
Carboxyfluorescein diacetate (CFDA)
Carboxyfluorescein succinimidyl ester (CFSE)
Cell blood count (CBC)
Cerebro-spinal fluid (CSF)
Chronic lymphocytic leukemia (CLL)
Citolytic T lymphocyte (CTL)
Coefficient of variation (CV)
Count of Blood cells (CBC)
Dermal DC like DC (iDC)
Digital signal processing (DSP)
Dimethysulfoxide (DMSO)
Dithiotreitol (DTT
Electron microscopy (EM)
Fluorecein isothyocyanate (FITC)
Fluorescence in situ hybridization (FISH)
Forward scatter (FSC)
Green fluorescent protein (GFP):
Herpes simplex virus type 1 (HSV-1)
High Resolution cell sorter (HiReCS)
Human dendritic cell line (HDCL)
Human immunodeficiency virus (HIV)
Immune mediated thrombocytopenia (IMT)
Interferon alpha (IFN α)
Interferon gamma (IFN-γ)
Langerhans cell (LC)

Lawrence Livermore National Laboratory (LLNL)
Macintosh (Mac)
Mean fluorescence intensity (MFI)
Monoclonal antibody (mAb)
Multi carousel loader (MCL)
Multiplicity of infection (moi)
N-ethyl maleimide (NEM)
Non-immune thrombocytopenia (NIT)
Ovoalbumin (OVA)
Paraformaldehyde (PFM)
Peridinin chlorophyll protein (PerCP)
Peripheral blood mononuclear cells (PBMC)
Phosphokinase C (PKC)
Phycoerythrin (PE)
Pisum sativum aglutinin (PSA)
Platelet-associated immunoglobuling (PaIg)
Polymerase chain reaction (PCR)
Propidium iodide (PI)
Red blood cells (RBC)
Reticulocyte (RTC)
Rhodamine 123 (R123)
Side scater (SSC)
Succinimidyl trans-4-(N-maleimidylmethyl)cyclohexane-1-carboxylate (SMCC)
White blood cells (WBC)
X-Ypeak (XYp)

Index

β-gal: 37, 77, 81
2nd Ab: 73
3-(2-pyridyldithio)propionic acid N-hydroxysuccimidine ester (SPDP): 90, 91
325 nm: 10
42D1: 73
4'-6-diamidino-2-phenylindole (DAPI): 59
488 nm: 5, 7, 10, 58, 63, 71
4Way Sorting: 14
5-, 6-carboxyfluorescein diacetate succinimidyl ester (CFSE): 80, 86. See CFSE
518 nm: 58
635 nm: 5, 7
650 nm: 5
7H5: 73, 76

A

AccuSort: 13
Acquired Immunodeficiency: 47. See AIDS
Acquisition: 6
Acridine orange: 60
Acrosomal integrity: 57
Acrosome: 57
Actin: 3
Activation markers: 49

Adobe PhotoShop: 39
AIDS: 2, 84
Air-cooled argon-ion laser: 15
Air-cooled laser: 12, 13
Allophycocyanin (APC): 44, 78, 85
Allopurinol: 63, 64, 65, 66, 84
Altra: 13
AltraSort: 13
Amphotericin: 63
Amplicon vectors: 38
Amplicon: 37, 67, 68, 77, 79, 81, 86, 87. See HSV-1. See also Vectors
Amplitude: 29
Analysis region: 18
Analysis station: 8
Analytical precision: 1
Analyzer(s): 10, 12, 41
Aneuploid(y): 30, 34. See DNA
Aneuploid/euploid DNA content ratio: 34. See DNA
Antibodies: 42, 43, 53, 78. See mAb. See also Monoclonal antibodies
Antigen uptake: 75, 78
Apoptosis: 77
Argon laser: 5
Argon-ion laser: 10, 15
Atopy: 50
Autoclone: 13
Autoimmune diseases: 45
Autoradiograms: 31

B

B220/CD45R: 74
Baboon: 95
BAC: 69
Batching: 24, 26
BD LSR: 10
BD: 10, 11, 17, 31, 55, 63, 71, 81
Beams: 14
Beckman Coulter: 12
Becton Dickinson (BDIS) (BD): 10, 11, 17, 31, 55, 63, 71, 81. See BD
Benchtop flow cytometers: 5
Benchtop: 5, 8, 10, 41
BG-6: 73, 77
Biohazardous: 8, 11
Biosafety: 53
Bivariate BrdU/DNA analysis: 29, 31
Bivariate cytokeratin/DNA analysis: 29
Blast: 18, 41
Blue laser: 7
Boars: 57, 58, 59
Bone marrow: 18, 41, 44, 48, 78
Bovine: 55, 59
BrdUrd: 31
Breast tumors: 33
BU10: 73
Bulls: 57, 58, 59

C

Calf thymocyte 60
Calibrite: 7
Cancer: 83
Canine: 643
Carboxyfluorescein diacetate (CFDA): 57, 58
Carboxyfluorescein succinimidyl ester (CFSE): 63, 64, 65, 66, 76, 84, 89, 90. See CFSE
Catcher tube: 11
Cauda epididymis: 60
CD11a: 49, 50
CD16: 47
CD19: 44, 47
CD23: 44
CD25: 49
CD27: 49
CD28: 49
CD3: 47, 48
CD34 (stem cells): 48, 49
CD38: 49
CD4 enumeration: 47
CD4 lymphocyte: 2
CD4: 2, 17, 19, 20, 43, 47, 48, 50, 51, 74, 84
CD45: 17, 43, 44, 47, 48, 49, 74
CD45RA: 49, 50, 51
CD45RO: 49, 50, 51
CD5: 44
CD56: 47
CD57: 49
CD61: 46
CD61: 47
CD62L: 49, 50, 51
CD69: 49
CD71: 49

CD8: 17, 19, 20, 47, 51
CD83: 71
CD95: 49
CDNA: 38
CDw60: 49
Cell biology: 12
Cell blood count (CBC): 47
Cell cycle: 2, 12, 23, 26, 29, 30, 31, 34, 57, 61, 83. See DNA
Cell division: 65
Cell sorter: 11
Cell sorting: 10, 11, 37, 83
Cell toxicity: 66
CellFit: 31
CellQuest: 10, 17, 65, 72
Cellular autofluorescence: 48
Cellular biology: 37
Cellular membranes: 3
Centricon: 89, 90
Cerebro-spinal fluid (CSF): 41, 43
CFSE: 63, 64, 65, 66, 76, 80, 84, 86, 89, 90
Channel: 18
Chicken: 60
Chromatin structure: 60
Chromosome enumeration: 12
Chromosome painting: 39
Chromosome: 37, 38, 39, 69
Chronic lymphocytic leukemia (CLL): 41, 44
Cis-acting: 67
CMRF-44: 71, 73, 77
CMRF-56: 73, 76, 77
Cockerels: 59
Coefficient of variation (CV): 14, 19, 34. See CV
Compensation: 26, 78
Confocal microscopy: 37, 39
Contour plot: 17, 20, 22, 23, 24, 25, 26, 44
Cosmids: 67
Coulter EPICS XL: 12
Count of Blood cells (CBC): 45
CTL: 51, 77, 79, 80, 81, 86, 87
CV: 14, 34
CyClone: 14
Cytofluorometers: 2
Cytokines 48, 55, 74, 75
Cytology: 42
Cytolytic T lymphocyte (CTL): 51, 77, 79, 80, 81, 86, 87. See CTL
Cytomation: 13, 14
Cytometer: 6, 8, 13, 14, 15
Cytometry: 1
Cytotoxicity: 67, 68, 69, 70, 80, 81

D

DAPI: 10
Data analysis: 17
DC lines: 51
DC uptake: 86
DC vaccine: 79
DC: 20, 71, 72, 73, 75, 76, 78, 80, 81, 85, 87
DCGM-4: 73, 76, 77
DC-LAMP: 73, 76
Dead sperm: 59
DEC-205: 71
Dendritic cell: 2, 20, 21, 48, 51, 71. See DC
Density plot: 18, 20, 22, 23, 24, 25, 26, 61, 68, 72, 80

Dermal DC like DC (iDC): 71.
    See DC
Dermis: 78
Digital signal processing (DSP): 12
Dimethysulfoxide (DMSO): 89, 90
DIOC3: 79, 80, 81, 86
Diode laser: 5
Diploid: 30, 34, 60, 61
Dithiotreitol (DTT): 91
DNA content :33
DNA index: 30, 34
DNA measurement: 29
DNA probes: 39
DNA: 3, 11, 12, 15, 18, 23, 26, 29, 30, 31, 32, 34, 39, 53, 58, 59, 60, 61, 65, 67, 69, 84
Dogs: 57, 63
Domestic animals: 58, 59
Dot plot(s): 17, 18, 19, 20, 22, 23, 24, 25, 26, 32, 34
Doublet discrimination module: 30
Doublets: 29, 30, 32, 61
Droplet: 3, 14
Dual parameter: 57, 60
Dual staining 61, 65
Dyes: 7, 10, 29, 30, 44, 48, 53

E

E. coli: 69
Effusions: 42
Electron microscopy (EM): 77
ELISA: 55
Enumeration of stem cells: 48

EPICS ALTRA: 12
Epidermis: 78
Erythrocytes: 60, 93, 95
EXPOTM: 13

F

FACS analysis: 9
FACS Comp: 7
FACS Vantage: 11
FACS: 79
FACSCalibur: 5, 8, 9, 10, 11, 53, 63, 71, 81
Fc binding: 48
Feline: 22, 23, 24, 25
FITC: 7 , 17, 20, 21, 44, 46, 63, 72, 75, 78, 85
FL1: 7, 72, 80
FL2: 7, 30, 34, 86
FL2A (pulse-area): 30, 32, 33, 34
FL2H (FL2 high): 30
FL2W (pulse-width): 30, 32, 33, 34
FL3: 7, 86
Flexible sorting: 15
Flow cell: 6, 12, 13
Flow cytometer(s): 1, 3, 5, 9, 12, 23, 25, 30, 53, 60, 78, 83
Flow cytometry intruments: 5
Flow cytometry: 1, 2, 3, 83
Flow microfluorometry: 59
Flow solution: 6
FlowCentre: 13
FlowJo: 10, 23, 24

Fluorecein isothyocyanate (FITC): 7, 17, 20, 21, 44, 46, 63, 72, 75, 78, 85. See FITC
Fluorecent probes: 3
Fluorescence *in situ* hybridization (FISH): 37, 38, 39
Fluorescence intensity: 2
Fluorescence: 1, 18, 22
Fluorescent antibodies: 2
Fluorochrome: 5, 7, 11, 18
Fluorogenic: 7
Forward scatter (FSC): 2, 5, 17, 20, 21, 26, 45, 46, 71, 73, 80. See FSC
Fotomultiplier: 6
F-plasmid: 69
Fragmentation: 77
Free calcium: 3
Free faty acids: 3
FSC: 2, 5, 17, 20, 21, 26, 45, 46, 71, 73, 80

G

G0/G1/S/G2/M: 32, 33
G0/G1: 32, 33, 34, 61
G0/G1+S+G2/M: 61
G1-phase: 30, 32, 33
G2/M-phase: 30, 32, 33
G2+M: 61
G2-phase: 30, 33
Gametes: 59
Gate: 17, 20, 21, 25, 26, 44, 47, 61, 72
Gating: 17, 23, 34

Gene therapy: 85
Gene transfer: 67, 69
Genetics: 12
Genitourinary tissue: 78
Genome: 67
GM-CSF: 71, 76, 80, 86
Granularity: 2, 45
Granulocyte: 2
Green chanel: 7
Green fluorescent protein (GFP): 37, 68, 69

H

Handling of samples: 53
Haploid: 60
HDCL: 76
He-Cd laser: 10
Helper virus: 67
Hematology counters: 5
Hematology: 42
Hematopoietic neoplasm: 33
Herpes simplex virus type 1 (HSV-1): 38, 67, 77
High Resolution cell sorter (HiReCS): 14, 15
High speed sorting: 13
High Speed: 14
High throughput: 1
High-speed sorter: 10, 11, 13
HIGHSPEED: 11
HISPEED: 15
Histogram overlays: 21
Histogram(s): 17, 18, 30, 31, 33, 34, 65, 72, 74, 75, 78

HLA-DR: 49
Hoechst 33342: 10, 59, 60
Homologous recombination: 67
Hormones: 3
Horses: 46, 57
HP: 25
HP-F1: 73, 76
HSV-1 gB: 77
HSV-1 immediate-early (IE) gene: 69
Human dendritic cell line (HDCL): 74, 75
Human genome project: 13
Human immunodeficiency virus (HIV): 47, 63
Human: 13, 55, 58, 61, 63, 71, 74, 75, 78, 95
HyperSort: 12, 13

Immunophenotype: 74
Immunophenotyping of DC: 71
Immunophenotyping: 22, 41, 44, 47, 48, 49, 53, 78
In situ PCR: 39
Inclusion bodies: 45
Indo-1: 10
Intel Pentium: 13
Interface: 39
Interferon alpha (IFN α): 55
Interferon gamma (IFN-γ): 50, 75
Internet: 10, 26
Intestine: 78
Intracellular antigens: 55
ISAC: 53
Isotype control: 20, 21, 22, 42, 74
Isotype matched: 78

I

IB4: 90
ICP27: 69
IDC: 72, 73, 74, 76
IgG: 45
IL-4: 50, 51, 71, 76, 80, 86
Immature reticulocyte fraction IRF: 44
Immune mediated thrombocytopenia (IMT): 45, 46
Immunize: 79
Immunobiology: 12
Immunodeficiency: 47
Immunofluorescence: 78

J

Jet-in-air: 11, 12

K

Ketoconazol: 63
Kolmogorov-Smirnov statistics: 18, 21

L

Langerhans cell (LC): 71, 73, 74, 76, 78
Laser: 3, 5, 10, 12, 13, 15, 71
Lawrence Livermore National Laboratory (LLNL): 13
LCmDC: 72
Leishmania infantum: 63
Leishmania: 84
Leishmaniasis: 63, 65, 66
Lepromatous leprosy: 50
Leucocyte: 2, 17, 18, 22, 29
Leukemia: 41, 43, 83
Leukocytes: 46, 93, 95
LIVE/DEAD: 81
Living sperm: 59
Lung: 78
Lupus erithematous: 45
Lymphocyte gate: 48
Lymphocytes: 2, 17, 22, 23, 24, 25, 44, 47, 55
Lymphoid tissue: 78
Lymphomas: 2, 22, 41, 42, 43, 83
Lymphoreticular neoplasm: 33

M

M1: 21
mAb: 5, 17, 20, 21, 31, 55, 71
mAb: 71, 72, 73, 74, 75, 77, 94
Mac1/CD11b: 74
Macintosh (Mac): 9, 10, 24, 25
Macrophages: 48
Magneto-optical: 9
Mammalian sperm: 57
Mammalian: 67
Marker: 18
MC57: 77, 81
Mean fluorescence intensity (MFI): 18, 47, 69, 73, 75, 80. See MFI
Mean: 18
Median: 18
MEDLINE: 83
Membrane potential: 3
Memory T cells: 49, 50
Men: 57, 58, 60
Metaphase: 39
Methanol: 90, 91
MFI: 18, 47, 69, 73, 75, 80
MHC Class II: 74
MHC: 78
Mice: 57, 58, 60, 79, 87
Microbiology: 12
Microinjection: 37
Micromanipulation: 37
Micromanipulator: 38
Migration. 78
Mitochondria function: 3, 57
Mitochondria: 3
Mitosis: 31
MO: 9
MoFlo BTA: 13
MoFlo MLS: 13, 14
MoFlo SX: 13
MoFlo: 13, 14
Molecular biology: 12, 37
Molecular cytometry: 37
Monoclonal antibody (mAb): 5, 17, 20, 21, 31, 55, 71. See mAb
Monocyte: 2, 17, 48, 55

Moribund sperm: 59
Motility: 59
Mouse DC line: 74
Mouse: 77
Movilization protocol: 48
M-phase (mitosis): 30, 33
MRNA: 3, 39
Mucous membranes: 78
Multi carousel loader (MCL): 12
Multicolor analysis: 41, 42, 47
Multicolor: 78
Multiplicity of infection (moi): 68, 69
Mx: 55

N

Naïve T cells: 49, 50
N-ethyl maleimide (NEM): 91
NK cell: 47
N-methyl-D-glucamine antimoniate: 63
Non-immune thrombocytopenia (NIT): 45, 46
Nuclear condensation: 77
Nuclei acid dye: 33
Nuclei: 33, 57, 58, 59, 60, 84
Nucleic acid: 60

O

Optical filters: 12
Optics: 7
Origin of replication: 67
Overlayed: 74, 75
Overlaying: 78
Overlays: 65
Ovine: 55
Ovoalbumin (OVA): 78, 85

P

Pac signal: 67
Packaging-defective: 67
PaIg: 18, 45, 46, 47
Pan B cells: 47
Paraformaldehyde (PFM): 92
Parasite: 63, 65
Parasitology: 66, 84
PBMC: 17, 18, 20, 22, 23, 55, 71, 76, 78
PBS :79, 89, 90, 91, 92, 94
PC: 10, 25, 26
PCR: 11, 15, 37, 38, 39
PE Applied Biosystems: 39
PE: 7, 44, 78, 85, 90, 91
Peak: 18, 19
PE-Cy5: 44
Pentamidine: 63
PerCP: 7, 78, 85
Peridinin chlorophyll protein (PerCP): 7, 78, 85. See PerCP

Peripheral blood mononuclear cells (PBMC): 17, 18, 20, 22, 23, 55, 71, 76, 78. See PBMC
Peripheral blood stem cells: 49
Peripheral blood: 41
Peritoneal centeses: 42
PH: 3
Phenotyping: 18
Phosphokinase C (PKC): 50
Phycoerythrin (PE): 7, 44, 78, 85, 90, 91. See PE
PI: 34, 53, 55, 58, 59, 61, 63, 64, 65, 66, 68, 69, 80 84
Pig: 95
Pisum sativum aglutinin (PSA): 57
Plant cell analysis: 12
Platelet studies: 12, 47
Platelet-associated immunoglobuling (PaIg): 18, 45, 46, 47. See PaIg
Platelets: 45, 47
Pleural effusions: 42
Ploid(y): 2, 18, 19, 34, 60, 83
Polymerase chain reaction (PCR): 11, 15, 37, 38, 39. See PCR
Polytyrene beads: 7
Present antigens: 51
ProCount: 48
Proliferation: 76
Promastigote: 63, 64, 65, 84
Propidium iodide (PI): 34, 53, 55, 58, 59, 61, 63, 64, 65, 66, 68, 69, 80, 84. See PI
Protein synthesis: 63
Protocols: 89
Purine: 63

Q

Quadrant marker: 19
Quadrant: 72

R

R1: 17, 20, 21, 34, 48
R2: 34
Rabbits: 58, 59, 60, 95
Rabies: 79, 81, 87
Rams: 58, 59
RAN retrovirus: 75, 76
Rare cell: 10, 13, 15
Rare event: 15
Rat: 95
Real time PCR: 39
Recombination: 69
Red blood cells (RBC): 46
Red blood cells-Chicken/trout: 29
Red diode laser: 7
Red light: 5
Red613: 7
Replication: 67
ReticOne: 12
Reticulocyte (RTC): 12, 18, 44, 45
RFD-1: 73
RFD-7: 73
Rhodamine 123 (R123): 57
Rhodamine: 53
RNA: 11, 15, 34, 60
RNAase: 34

S

S phase: 61
Sample: 8
Saponin: 93, 94
Scattered light: 2
Scattergram: 19
Sephadex: 89, 90
Sex preselection: 60
Side scater (SSC): 2, 5, 17, 20, 21, 34, 44, 45, 46, 48, 72, 80. See SSC
Single-cell gene transfer: 37
Single-cell sorting module: 11
Singlet(s): 29, 30, 61
Sodium hypochlorite: 53
Sodium stibogluconate: 63
Solid tumors: 33
Sort cells: 38
Sort modes: 13
Sorter 60
Sorting: 3, 8, 9, 11, 12, 13, 14, 15
Sortmaster: 14
SortSense: 12, 13
Spectra/Physics: 15
Sperm quality: 57, 61
Sperm: 9, 61, 57, 61, 65, 84
Spermatid: 60
Spermatogenesis: 60
Spermatozoa: 57, 60, 61, 84. See Sperm
S-phase (DNA synthesis): 30, 32, 33
SSC: 2, 5, 17, 20, 21, 34, 44, 45, 46, 48, 72, 80.
Staining: 53
Sterile cloning: 11
Sterile: 11
Streptavidin-PE: 47
Sub-platelet particles: 46
Subpopulations: 2
Subsets: 71
Succinimidyl trans-4-(N-maleimidylmethyl)cyclohexane-1-carboxylate (SMCC): 90, 91
Surface markers: 51, 71, 74, 77
Surface receptors: 2, 3
Surface staining: 41
Surface: 2
SYBR-14: 58, 59, 61, 63, 64, 65, 66, 84

T

T cells: 49, 50, 51, 80
T cytotoxic/suppressor: 47
T helper: 47
TaqMan: 37, 38, 39
Testicular: 60
TetraOne: 12
Tetraploid: 60
TGF-β1: 71, 76, 80. 86
Th1: 50
Th2: 50
Thermocycler: 39
Thiazole orange: 44, 45
Thrombocytopenia in horses: 46
Thrombocytopenia: 45, 46
TNFα: 75
Total IgG: 45, 46
TPD153: 73

Transduce cells: 37
Transduction efficiency: 67, 68, 69, 70
Transduction: 38
Transgene: 67
Transplantation: 48
TriColor: 7
Tuberculoid leprosy: 50

U

UV excitation: 10
UV: 10

V

Vaccine: 79
Vas deferens: 60
Vector: 67, 68, 69, 70, 85
Viability: 57, 60, 64, 65
Viable: 57
Virions: 67
Virus: 3, 69, 75

W

Waste: 6

Water cooled Spectra/Physics argon-ion laser: 15
Water quality: 12
Water-cooled laser: 13
Wavelenghts: 2
White blood cells (WBC): 45
WinMDI: 10, 26, 65

X

XL system II software: 12
X-spermatozoa: 58, 59, 60. See sperm
X-Ypeak (XYp): 59, 60. See sperm

Y

Y-spermatozoa: 58, 59, 60. See sperm

Z

Zeiss Axiovert digital microscopy: 39
ZM3.8: 73

Other Books of Interest

Development of Novel Antimicrobial Agents: Emerging Strategies
Editor: **Karl Lohner**, *Graz, Austria*.
International researchers from academia and industry present this unique collection of highly acclaimed reviews covering every aspect of this important topic. Essential reading for all scientists interested in the development of antimicrobial agents.
2001, 284 p. ISBN 1-898486-23-9 £84.99 or $169.99

Peptide Nucleic Acids: Protocols and Applications
Edited by: **Peter E. Nielsen** and **Michael Egholm**
Contains state-of-the-art protocols and applications on all aspects of PNA. Concepts are explained clearly and in practical terms and each chapter contains concise background information. Written by leading experts in the field.
1999, 262p. ISBN 1-898486-16-6 £59.99 or $119.99

NMR in Microbiology: Theory and Applications
Eds: **Jean-Noël Barbotin** and **Jean-Charles Portais**.
Foreword by **Daniel Thomas**
This book describes the theory and practical applications of this increasingly important technique and is aimed specifically at microbiologists. Discover the value and potential and of this powerful technology!
2000, 500 p. ISBN 1-898486-21-2 £84.99 or $169.99

Intracellular Ribozyme Applications Principles and Protocols
Edited by: **John J. Rossi** and **Larry Couture**.
Foreword by **Thomas R. Cech**.
The definitive guide, containing reviews of the most recent pharmaceutical, therapeutic, and biotechnological applications of ribozymes.
1999, 289 p. ISBN 1-898486-17-4 £74.99 or $149.99

Oral Bacterial Ecology: The Molecular Basis
Editors: **Howard K. Kuramitsu** and **Richard P. Ellen**
Aimed at researchers in the field of oral microbial pathogenesis this book reviews in depth the molecular basis of oral pathogenesis and ecology. Particular emphasis is placed on recent advances in molecular microbiology and genomics.
2000, 314 p. ISBN 1-898486-22-0 £74.99 or $149.99

H. pylori Molecular and Cellular Biology
Editors: **Mark Achtman** and **Sebastian Suerbaum**
Leading international scientists have contributed comprehensive and topical reviews. The book is essential reading for all researchers and clinicians working in the *H. pylori* and related fields.
2001, c. 350p. ISBN 1-898486-25-5 £tba or $tba

NO. OF COPIES	TITLE	PRICE
Add postage and handling £4 (UK), £6 (Europe), £8/$16 (Rest of World) for each book:		
	TOTAL	

☐ I enclose a cheque/check in £/$. Amount ....
☐ Please debit my Visa/Mastercard/Diners/Amex. Amount ....
Card number ....
Expiry date ....
Name:
Address:

Order from:
Horizon Scientific Press
32 Hewitts Lane, Wymondham
Norfolk, NR18 0JA, U.K.

Tel: +44-(0)1953-601106
Fax: +44-(0)1953-603068
Email to: mail@horizonpress.com
http://www.horizonpress.com

In the USA order books from: ISBS, 5804 N.E.
Hassalo St, Portland, Oregon 97213-3644
Tel: (800) 944-6190 Fax: (503) 280-8832